To Charlie
With Love

Gary

Christmas 2002

Farming Day by Day –
the 1960s

Farming Day by Day – the 1960s

As recorded in the *Daily Mail* by
John Winter

Selected and edited by
Peter Bullen, Phillip Sheppy and Derek Watson

Old Pond Publishing

ISBN 1 903366 21 6

Published by

Old Pond Publishing
104 Valley Road, Ipswich IP1 4PA, United Kingdom

www.oldpond.com

Typesetting by Galleon Photosetting
Cover design and illustrations by Liz Whatling
Printed and bound in Great Britain by
Biddles Ltd, Guildford & King's Lynn

Foreword by The Lord Plumb of Coleshill

We are privileged to have such a detailed account of John's articles covering more than a decade of farming day by day, and I am grateful for the opportunity to share, re-live and reflect on these memories through the 1960s.

During that time, farmers, traders and consumers were able to benefit from stories told by our teams of agricultural correspondents; stories about farming businesses, great or small, and individuals who were living through the struggle and progress of the industry. Many Birmingham businessmen would question me on the, sometimes provocative, daily reports which we enjoyed reading as we travelled to London on a Monday morning.

John's Lancastrian upbringing was reflected in the warmth of his articles about the families of the land, reminding us for instance in 1961, that a quarter of Britain's food was produced by people who were little more than peasants, sweating over their labour without thoughts and hopes of tomorrow! At that time there were 166,000 farmers farming under 20 acres. In the same article that year, Travers Legge, editor of *Farmers Weekly*, said "If we reach a stage at which the minimum would be 35 to 40 acres of better land, we could offer competition in a Common Market which *no* country in Europe could match!" Now it is around 200 and increasing.

John reminds us that, at that time, the minimum wage was £8.90 for a 46-hour week and the income of hill farmers in Wales was £250 a year. If he were writing today, he would reflect on the enormous change in farm structure and the drift from the land of some 40,000 farmers and farm workers over the past two years. And this is still continuing at an unbelievable rate.

His fascinating reports take us through the times of change when my predecessor as Chairman of the Animal Health Committee of the NFU, Salter Chalker, fought and succeeded in getting an eradication scheme operating to eliminate tuberculosis. As TB was adding risk to the consumer, 24,000 cattle infected by the disease were slaughtered every year. It is sad that now, in 2002, we are seeing a return of outbreaks which is believed to be spread by badgers.

Stockbreeders, however, were beginning to look abroad. We believed we were the livestock yard of the world with the finest breeds, but on 1st November 1961, John reported the import of 30 Charollais bulls from France with a disease security curtain drawn around them so tightly that only authorised handlers would see their arrival. Three were found to have blood infection and shot; others were sent to AI stations never to have direct contact with cows, but the disease in question was not found transmissible through semen.

This leads to the big political issue of the early 1960s, which was of course the Common Market. On 16th January 1962 John reports "Farmers say 'no' to the Six: Common Market levy would raise prices and hit the Commonwealth." Harold Woolley, then President of the NFU, was reported by John from the 1962 Annual Meeting in London, saying "it would mean dearer food, higher costs for our export trades and social consequences which would bear heavily on people with fixed incomes." The glasshouse growers and market gardeners really believed that they would be doomed.

Whilst it was wrong to suggest that Britain was being kept out because of agriculture or horticulture, with hindsight, had we joined the original Six in the early 1960s, we could have helped shape events rather than react, giving us a much stronger voice. On 15th November 1962, John reported that there was not a single dissentient voice raised against Ted Heath when he spoke to 1,500 farmers giving a progress report on Common Market negotiations. His speech was received with a big ovation, there was no trace of hostility and some were even enthusiastic for the European link-up.

As other Continental breeds of cattle, sheep and pigs were gradually imported, the shape and colour of our cattle across the hills and valleys changed as their popularity increased. They are now established amongst our good British breeds and can provide a more than adequate supply of home-grown meat. However, following one of the worst years on record in farming where herds and flocks have been wiped out through foot-and-mouth disease, security of imports should be a first priority. Apart from an unknown quantity of meat products being illegally imported into the UK, 180,339 tonnes of meat were legally imported during the recent outbreak from six countries where foot-and-mouth is endemic. No other country in the world would allow this to happen!

All this, presumably inspired by interest in a cheap food policy, is no less than an act of committing economic suicide.

Again, if John were reporting on farming issues today and had the freedom to put his thoughts on paper, he would be very critical of the red tape and bureaucracy related to food and farming. He would be explaining and commenting on the business of diversification and multi-functionality. I believe he would be sympathetic towards modulation since he would recognize a way of helping family farmers to stay in business as subsidies are reduced. He would be following the development of organic farming and biotechnology. He could justifiably look back at his comments on 28th May 1964, when Government announced a scheme which he called "Big help for the little man". "The choice", he said, "was factory or family farming."

He would today say that very few took advantage of such a financial incentive while they enjoyed guaranteed prices and deficiency payments, yet they watched the supermarkets grow. One of the lessons we did not learn is that the UK has been, and still is, the least co-operative country in the European Union and we now pay the price.

But above all, his pen would be reflecting his sincerity, his Christian belief, and his loyal support to the British farmer.

During my time as an office holder and President of the NFU, John became a very good friend. This book truly reflects his faithful reporting during his time with the *Daily Mail*. It will serve not only as a book of memories but also as a valuable source of reference covering a fascinating period of farming history.

Editorial Note

John Winter was careful to safeguard all the reports on farming and food matters published in the *Daily Mail.* Covering the period 1959 to 1979, John kept one copy pasted into a scrapbook and a second filed into a loose folder. After his death in 1980, his widow Phyllis passed both sets of cuttings to Derek Watson with the request that he deal with them as he felt appropriate.

Peter Bell, former president, chairman and secretary of the Guild of Agricultural Journalists, based a series of articles on the reports which were published in the Farmers' Club *Journal.* The loose cuttings were donated to the Rural History Centre at Reading University and the scrapbooks to the Royal Agricultural Society of England (RASE) at Stoneleigh.

There matters rested until 1999 when the RASE's Honorary Librarian, Phillip Sheppy, suggested to Derek Watson that efforts be made to find a publisher and they jointly approached Old Pond Publishing. Peter Bullen, assistant to John Winter from 1963 to 1971, joined Phillip and Derek to form an editorial team.

Using the RASE scrapbooks, each member of the team took responsibility for making a selection for particular years, aiming to reflect fairly the main issues and the 'flavour' of the period. The final selection of nearly 236,000 words represents a mere fraction of John's output.

This volume presents the first half of the selection, ending in 1969. As the reports speak for themselves, they have been published without editorial comment or explanation. They have been edited only to remove sub-heads and some small secondary items and to join the previously very short paragraphs.

It is hoped to publish the second half of the selection, dealing with the 1970s, in the not-too-distant future.

Peter Bullen, Phillip Sheppy,
Derek Watson, April 2002

Acknowledgements

The editorial team and publishers wish to thank the following:

Associated Newspapers, for permission to publish material which originally appeared in the *Daily Mail*.

The Royal Agricultural Society of England for its support in the preparation of this material.

The Guild of Agricultural Journalists for support.

Halina Bannister for typing.

Sue Gibbard for indexing.

We are grateful to the following for permission to reproduce photographs: Museum of East Anglian Life, Stowmarket; Rural History Centre, University of Reading; Universal Pictorial Press and Agency Ltd.

Farm Mail from the Inside

Peter Bullen

Farm Mail was launched in 1961, a couple of years after John Winter had left Manchester to join the head office staff of the *Daily Mail* in London as the paper's agricultural correspondent.

Born in Bury, he was a Lancastrian to the core and proud of it. He often caused consternation when dining in hotels and restaurants in more southern counties by demanding cheese should be served with his apple tart and by growling loudly "The Duke of Lancaster" (Her Majesty's full title according to him and other Lancastrians) whenever the toast to "The Queen" was proposed at formal dinners.

His full name was really Winterbottom, a popular name in his native county, which he swore was shortened to Winter to make a more compact by-line in the early years of his career when newsprint was severely rationed. After years of use it was too late to go back to the full version when newspapers were eventually allowed unlimited pages.

The launch of Farm Mail was a significant move by the *Daily Mail* to establish itself as a clear leader in the regular reporting of news of Britain's largest and, many argued, most important industry. It was an inspired move as once it had been announced it effectively prevented any other national daily paper doing the same without running the risk of being accused of plagiarism.

The move has also to be seen against the background of that time in the late 50s and early 60s which, in retrospect, was the golden era for agricultural reporting in the national press. Every quality daily had a full-time agricultural correspondent on the staff, some even more.

The Times had Leonard Amey assisted by Sam Platts; The *Daily Telegraph*, W D Thomas; The *Guardian*, Stanley Baker; while the *Mail*'s deadly rival, Lord Beaverbrook's *Daily Express*, had Alexander Kenworthy, assisted by Edward Trow and in Manchester, John Scholes. For its part, the *Daily Mail* had John Winter and in its Scottish office, Roy Gregor.

I had been working on Lord Beaverbrook's weekly agricultural paper, *Farming Express*, which he had been publishing to spearhead his campaign against entry into the Common Market. When President Charles de Gaulle vetoed Britain's entry Lord Beaverbrook sold *Farming Express* in 1963.

For me, the timing was fortuitous because John Winter wanted, in fact needed, an assistant following heart trouble which had been exacerbated by pressures of work. So I crossed from one side of Fleet Street to the other to work for Lord Rothermere's *Daily Mail* as assistant agricultural correspondent.

The creation of the daily Farm Mail column had itself increased the pressure on John Winter as it appeared six days a week, throughout the year. Even the weekend was not free as the Farm Mail piece had to be in Monday morning's paper. This meant working on Sunday, or working twice as hard on Friday to write Saturday morning's piece and another to be left clearly marked "Sunday for Monday".

With two of us working it did become easier to arrange to have alternate weekends free. However, journalism, particularly on a daily paper, is a 24-hours-a-day business that does not lend itself to time off, so completely clear days off were almost non-existent.

This was particularly true in John's case as he was renowned for his constant telephoning around his wide circle of contacts, not only from the office in working hours but before eight in the morning, late into the evening and on Saturdays and Sundays at all hours. He was known to a few close friends as "Jumbo", as much for his many trunk calls as for his prominent ears.

He, like all of us writing about farming in those pre-mobile-phone days, had earmarked public telephone boxes and AA/RAC call boxes across Britain to use to dictate stories to his paper as he travelled to and from shows, press conferences and other events. It was a standing joke among other agricultural journalists that, if they were driving in some remote part of Britain and saw a lone car parked next to a telephone box on the crest of a windswept hill, closer inspection would reveal that it was John's car and he was on the phone again. And that did happen more than once!

If there was one thing John liked especially about the job it was travelling around the country covering the large agricultural shows, visiting farms and doing roundups of harvesting progress. Whereas we both attended the national shows like the Royal Show at Stoneleigh, the Royal Dairy and Royal Smithfield Shows held at Olympia and Earls Court in London, John covered the large regional shows by himself while I remained in London writing the daily Farm Mail pieces.

On these occasions he may have been away from the pressures of Fleet Street itself but covering the shows was no easy option. I know, I did some myself. For each show the *Daily Mail* would produce what is called a slip page – a special edition of several thousand copies that went on sale in and around the showground for the two or three days of the event as well as being sold in the surrounding towns and villages.

This meant filling a half to a full page (broadsheet size not tabloid) of news from the show plus all major results which had to be found, gathered, written and telephoned by late afternoon, as slip pages usually went in the paper's early editions that had to be printed and rushed to the far corners of the country before morning.

Often, both John and I might be out of London covering important agricultural stories. If a major story then erupted in London perhaps on the Press Association or Reuters tapes we would have to react to it wherever we were, hopefully from a hotel lobby or room but often in an isolated telephone box in the depths of the countryside.

It usually meant ringing some trusted contacts in the National Farmers' Union, the Ministry of Agriculture or other organisations, or else providing some sort of commentary from one's own knowledge of the topic. That is why, like all farm writers on national papers, we carried briefcases bulging with relevant statistics and clippings of stories likely to re-occur such as disease outbreaks, export or production figures or pronouncements by industry leaders and Government ministers.

Too often for our liking we would learn that our really good lead story to the Farm Mail had been selected for the *Daily Mail*'s front page or another leading news page – and we would be told to find another Farm Mail lead out of thin air. Not always an easy task!

In addition to the daily Farm Mail items we would regularly be asked to write features on current food and farming topics and cover major news stories destined for the front page. Four or five times in most years we were proud to see we had written the main front page "splash" story.

When early in my career on the *Daily Mail* I got my first front page "splash" – a big event in any journalist's life – I received a telegram of congratulations. It was from John who was at a farm show in the west of England. This was typical of him but most untypical of the usual relationship between specialist writers and their assistants in the cut-throat environment of Fleet Street papers in those days.

But ours was always a friendly relationship built on mutual trust. We lived a few miles from each other in Surrey and mostly travelled to work and back on the same train and bus. John Winter was not the stereotypical hard-bitten, hard-drinking Fleet Street journalist. True, he enjoyed an occasional drink and he did smoke small cheroots or his pipe most of the day, but he was a kind-hearted family man – and the only Fleet Street journalist

I know who regularly read his Bible before going to sleep at night. If he had a vice, it was that he was undoubtedly a workaholic.

Our working relationship would probably have continued undisturbed way past its eighth year had not the *Daily Mail* taken over its stable mate the *Daily Sketch* in 1971 and eventually become a tabloid newspaper. The merger resulted in 50 per cent of the journalists being made redundant, including me.

Until then, a normal working day for us would be getting into the office between 10 and 10.30 am, having avidly read the *Mail* and several rival papers on the train on the way in. The stories we found for Farm Mail came from many different sources: events such as press conferences, Government publications and announcements, agency messages, tips or stories sent in by local correspondents in the counties, items gleaned from other publications, scientific reports and often just from conversations with contacts throughout the industry.

Occasionally a whole day might be spent at the office chasing up various ideas and leads for stories. At other times the morning could be spent at a Ministry press conference followed by a formal lunch in the City and a personal interview with an industry leader or politician in the afternoon.

Somehow the Farm Mail piece or pieces would have to be written and submitted by about 6 pm. If you were lucky you could be home by 8 pm and just going off to sleep about midnight when the phone would ring and your news desk would be asking if you knew such and such a story was on the front page of a rival's first edition. That would mean checking the story out by ringing a contact or two when they too were probably asleep in bed. You had to have good contacts to do that!

In selecting or writing stories for the Farm Mail, or anywhere else in the paper, we aimed to produce items that literally anyone would find interesting. It meant reducing often very specialised data into simple, readable copy that would not only appeal to the millions of lay readers of the *Daily Mail* but would also be read by the specialist farmer or agriculturalist.

The following pages will hopefully indicate the success of the technique.

1959–1961

In this first period of Farm Mail *the articles were shorter and less frequent than they were to become in subsequent years. Items were of a more general interest, among them the saga surrounding marketing boards, including the battles of "rebel" Jack Merricks. This was a time of growth in the further education programmes at county level and in the important work of the Young Farmers' Clubs. Disease problems meant that the first import by MAFF of Charolais (always spelled by John with a double 'l') cattle blood into this country all but ended in disaster. At long last, 1960 saw Britain's farms cleared of bovine TB.*

 ## POLITICS and FOOD

Shops get 'butter spivs' warning

John Winter 11.11.59

A warning against butter racketeers who are trying to create a Black Market by asking 1s a lb above normal prices went yesterday to 11,000 small grocers. It came from Mr Tom Lynch, president of the Small Shopkeepers' Union, after a Midlands wholesaler rejected an offer of a "substantial" supply of butter at the Black Market price.

Mr Lynch said: "When the spivs realise they are getting nowhere with the big men they will switch their attention to small traders. Our members are being told to pay fancy prices. If they do they will have to pass on the increase to the public, and our policy is to keep prices down."

The spivs have also approached members of the National Federation of Wholesale Grocers. All the offers were turned down. This was revealed by members of the federation's executive, in London for a conference.

The 6*d* milk machines start row

John Winter 16.11.59

Dairymen are protesting that sixpence-in-the-slot milk machines are stealing their trade. The complaint, made to the Milk Marketing Board by North-East milkmen, is the first grouse against the machines, which appeared in Britain two years ago.

Said Mr Patrick O'Neill, the dairymen's secretary, "Although the board is ready to supply any wholesaler with cartoned milk, consumers are inevitably switching from one supplier to another."

Egg men fear a shortage: prices may soar in the summer

John Winter 18.1.60

A serious shortage of eggs in the summer or autumn is feared by the Egg

Marketing Board. The shortage will be caused by farmers who, alarmed at the current low prices for eggs, are cutting back their orders for chicks. As a result, the Board believes there might be millions fewer hens laying this year. This would send up the prices of English eggs and leave the home market wide open to imports, especially from Denmark.

A hastily called meeting will be held this week between Board officials and leading chick hatchers to discuss the shortage fears. The Board will ask the hatchery men to declare figures from their order books. They are believed to be more than 20 per cent down on last year.

The hatchers, in turn, are likely to ask the Board to issue a plain warning to farmers about the prospective dangerous situation. They still have time, by booking immediate orders, to restore this year's laying flock to a reasonable level. But nothing can now be done to prevent a shortage of large eggs in September and October. The chicks to lay these should already have been hatched.

Current prices to farmers for large eggs – 3s 3d a dozen – will take a further dive this week to 3s. Standards will fall to 2s 9d and mediums to 2s 6d. At these prices, they say, production is not economic.

Cut costs bulk buying starts farm trade war   *John Winter 7.3.60*

A farm trade war threatens to spread throughout Britain. Already orders worth thousands of pounds a month, which farmers usually place with merchants, are being diverted direct to manufacturers. The farmers, organised in buying groups, are saving handsome discounts, but merchants losing the business are

threatening reprisals.

The cut-price buying movement, started last year by Mr J Rhys Thomas, former vice-president of the National Farmers' Union, has spread to 11 groups covering all Herefordshire. Their turnover this year could be well over £100,000. New groups are being formed in neighbouring counties, and requests for advice on launching more are pouring in from all over the country.

Group buying is a "dynamic" topic in every cattle market and village local. It cuts across old loyalties and is straining lifelong friendships. "The farmer has always bought retail and sold wholesale," said Mr Thomas. "Unless he can change to a sensible pattern of co-operative buying and selling, farming as we know it is sunk."

One group has saved an average of 17½ per cent on its purchases, and some discounts have reached 38 per cent. Members meet each month to pool their orders, and pay the next month.

Housewives can buy *Daily Mail* beef today   *John Winter 18.5.60*

The prize bullock reared on the Bovril Farm at Ampthill Park, Bedfordshire, in the *Daily Mail* Beef Project was inspected by 40 members of the Industrial Caterers' Association at Letchworth, Hertfordshire, last night. The carcass was expertly cut up by master butcher Mr Tom Dennis during a demonstration in the canteen of an electronic computer factory.

Mr Dennis bought the carcass – the only one of the six steers raised in the project to receive an "S" (Super grading) – from the slaughterers, the Fatstock Marketing Corporation, on Monday. He showed members of the Luton, Bedford, and District branch of the Industrial Caterers' Association – mostly managers and manageresses of large factory canteens – how to cut up the beef into joints and which points to look for.

The *Daily Mail* carcass was an excellent example for the demonstration, he said. "At 17 months this bullock has done amazingly well and is just what the butcher and the housewife wants," was Mr Dennis's verdict. The beef will be on sale today at Mr Dennis's shop in Ashwell, Hertfordshire.

Soames keeps farmers waiting

John Winter 3.8.60

If Mr Christopher Soames, the new Minister of Agriculture, has any new ideas for restoring farmers' confidence in the Government, farmers will have to wait for them. Yesterday, his first working day in his new job, he told a Press conference: "I am coming in at a challenging time. Farmers' efficiency has increased enormously, and production has gone up. But there are many problems, and the main ones are closely interwoven with national and international policies."

Mr Soames, at 39 the youngest member of the new Cabinet, has no intention of jumping into the stormy arguments about surplus production or the Anglo-Danish bacon war until he has examined them very closely.

He starts with a sound country and farming background. He was born on his father's farm in Buckinghamshire. For nearly ten years, after marrying Sir Winston Churchill's youngest daughter, Mary, he managed his father-in-law's 500-acre farm at Chartwell. As MP for Bedford for ten years, Mr Soames has also been actively interested in the diversified farming in his constituency. He has visited farms and attended National Farmers' Union meetings. "I have had the most cordial relations, and I know their problems," he said.

He has high hopes of success from talks in progress between the Government and the NFU, which were born out of the farmers' anger and frustration at the last Price Review, and launched by Mr Macmillan himself.

The shrinking egg – still the prices go up

John Winter 17.10.60

Dearer eggs are on the way, following a steady fall in home production. Supplies dropped by more than 2 million last week and large eggs will be unobtainable in many shops this week. Their price, which reached sixpence in some areas at the weekend, is likely to rise higher. Prices paid to producers at packing stations for all sizes of eggs go up today. Large and standard rise a penny a dozen, medium and small twopence.

The shortage is caused by heavy cuts last winter in farmers' orders for this year's laying birds, following overproduction and low prices last year. It was aggravated by a cut in the Government's guaranteed price for eggs, and heavy losses of laying birds from fowl pest. Supplies will continue to fall until the rise in production from this year's depleted pullet flock offsets the decline from older birds. This may take several weeks, and even then the rising production will be mostly in smaller eggs. Although farmers stepped up their chick orders on the advice of the Egg Marketing Board, in spring, these late-hatched birds take longer to lay their first eggs, and many will not be laying fully until nearly Christmas.

"It is difficult to forecast the turn in

the production curve, but we hope it will be showing by the end of November," said Mr C J Harrisson, vice-chairman of the board, yesterday. "The percentage of large eggs is a little above average, but, as total production is down, the shortage is quite sizeable. I blame the present position about 50 per cent on the Government, 20 per cent on fowl pest, and the rest on last year's low prices."

Imports will fill the gap of 15 million eggs a week. Supplies, which were disorganised by the London docks hold-up, are expected to pour in this week.

No bread can slim warns report

John Winter 18.11.60

Advertisements aimed at weight-conscious women, saying that special kinds of bread are slimming will be illegal if the Government accepts a food standards report published yesterday.

No bread, or any other food, can properly be called "slimming"; the Food Standards Committee on bread and flour say it agrees on this with the Joint Nutrition Panel. The committee also backs its recommendation that advertisements which claim weight-reducing properties for particular types of bread should be banned. The ban should also cover biscuits, rusks, rolls and certain breakfast cereals. Slimming claims, says the committee, usually rest on the ground that incorporation of a much larger amount of dry gluten results in a very considerable decrease in density, and therefore each slice contributes fewer calories to the diet.

"It is contended that bread is normally eaten by volume, and not by weight, but this is not necessarily true. No bread has any specific weight-reducing properties, and slimming claims, either made directly or by implication, can only mislead the consumer."

The committee accepts that incorporation of additional gluten (crude wheat protein) in flour, at the rate of 14lb dry material to a sack, produces bread of more open texture which is said to get stale less readily. People should be free to buy this kind of bread if they want to. But in stipulating a suitable distinguishing name the committee runs into difficulties.

The simplest course, it says, would be to call it after its most characteristic ingredient, but "gluten bread", it admits, is unattractive. Flours which contain 16–22 per cent protein could be described as "gluten bread" without any reference to protein. Those with less than 16 per cent should not qualify for special description. Bread with more than 22 per cent could be described as "high protein".

It's a turkey sell-out and geese and ducks are vanishing fast

John Winter 22.12.60

If you have not yet bought your Christmas turkey, goose, duck or chicken, do not delay in the hope of getting it cheap – or you may not get one at all. London's Smithfield wholesale market was cleaned out yesterday. And there are very few more birds to come. Turkeys hanging on stalls, or stacked, frozen, in cartons, were being held against order. There was not a goose or duck in sight. Retailers have birds on offer, but their repeat orders to Smithfield show that they are going fast too.

Prices have not rocketed, except for late, small deliveries of fresh turkeys, both English and Irish. They fetched up to 5s 6d a lb wholesale, for the popular 10–12lb birds. Fresh birds are, on average, 6d a lb dearer than last year. Frozen ones are 3d cheaper. Any rush for the last few birds this morning may produce fancy prices.

"We have never sold so much poultry at Christmas, and it looks as though supplies will only just meet demand," said Mr Leonard A Juniper, chairman of the Wholesale Poultry Merchants' Federation. "We cannot even meet the demand for pheasants, and when the wholesale trade finishes tomorrow I don't think there will be a bird left in cold store."

The British Turkey Federation estimates that 4.5 million turkeys, more than half of them frozen, have gone into the Christmas eating spree – 500,000 more than last year.

Greengrocers, their shops stacked with fruit, vegetables, nuts, holly and mistletoe, are waiting for the buying rush, which is usually delayed until the last two days before Christmas. Most prices are down on last year.

Missing – the 'top of the milk' loses its cream   *John Winter 22.12.60*

Thousands of housewives are looking at their milk with a sour eye these mornings . . . because the cream line at the top of the bottle has vanished. The only milk they can now get in half-pint bottles from their roundsman in some London areas is homogenised – so that the cream is mixed evenly through the bottle.

The change follows the disappearance of farthings as legal tender. From January 1 milk retailers have been permitted to round off the price of a half pint which contains an odd farthing, to the next halfpenny above. The order also raised the price of homogenised milk from 8¼d to 8½d a pint, and United Dairies decided they would supply only homogenised milk in half pints.

Typical of the complaints that followed was that by a Kensington woman: "I rang the dairy and the manager told me the 'creamless' milk was better, but I don't think so. It seems thin and watery and, in any case, I like to use the cream for soups and other purposes."

A United Dairies spokesman said the change was made on economic grounds. "Homogenised milk is increasing in popularity. It is smoother than other milk, and I think customers who have not had it before will prefer it when they get used to it," he said.

Farmers shun Common Market
John Winter 24.1.61

It would be a national disgrace if Britain sought entry to the European Common Market at the expense of the Commonwealth, said Mr Harold Woolley, the farmers' leader, yesterday. It would also mean dearer food, higher costs for our export trades and social consequences which would bear heavily on people with lower and fixed incomes.

Mr Woolley, president of the National Farmers' Union was speaking to 560 delegates at the union's annual meeting in London. "Half our trade is with the Commonwealth against 15 per cent with Common Market countries," he said. It was wrong to suggest Britain was being kept out of the Common Market because of her agriculture. The union believed the right course was to seek a gradual development of international integration.

Mr R N Wheelock (Monmouthshire) asked for a bold statement that the British farmer could confidently compete on equal terms with any farmer in Europe. Mr Woolley replied: "I am quite sure that we can hold our own, but we have to realise that entry into the Common Market would be a very serious change."

The meeting agreed to press for more rapid Government action under the anti-dumping laws. "There is little doubt that recent imports of Russian barley, American frozen and dried peas, Polish eggs and Argentine beef have damaged our industry, and there has been an element of dumping in all cases," said Mr W A G Kendall (Bedfordshire and Huntingdonshire).

No-milk-on-Sunday homes get gadget *John Winter 20.2.61*

Plastic "milk keepers" are to be offered by Express Dairies to all their 4,500 customers in Berkhamsted, Hertfordshire, where the company plans to end Sunday milk deliveries after this weekend. The "keepers" are in two parts, which fit together to form a rigid case to enclose a pint bottle. They are expected to be ready before the warm weather comes.

"This is an idea to help customers who do not have refrigerators," said a spokesman.

Berkhamsted Co-operative Society, main competition of Express, is not going to cut out Sunday deliveries. Said Mr H Johnson, manager: "We do not propose to canvass for extra trade, but, of course, shall not refuse to accept any customers who may wish to switch to us for a seven-day service."

If they have to become Common Market gardeners . . . 'We'll be doomed without government help', say Britain's glasshouse men *John Winter 21.6.61*

The June sun danced like liquid gold on the glass roof of London's market garden as the trains from Liverpool Street roared through the Lea Valley in Hertfordshire yesterday. Beneath the glass a harvest of more solid gold was ripening in the pungent humidity of tomato-houses. This could be the best season English tomato-growers have had for years, but it is more likely to be the worst – not through poor crops, low prices or because fickle customers develop a sudden aversion to tomatoes.

Over those fragile roofs hangs a shadow which could herald their extinction within four years – the shadow of the European Common Market. Most of the growers in the industry are intelligent. They accept that if it is in the major interests of Britain to enter the Common Market their own interests will have to give way. But if they are to get a body blow against which they can do nothing to protect themselves they at least expect the Government to recognise their plight and help them meet the tough competition they will face.

What is the horticulture industry? Its two basic crops are apples and tomatoes. It is an industry which provides over £200 million of the housewife's greengrocery purchases every year. The threat to the glasshouse producers comes from the highly developed Dutch growers; to the orchard men from the rapidly expanding fruit industry of Italy. Both home products are at present protected – tomatoes have an import tariff starting at 6d a lb and falling to 2d during the English production season. Apples come in on quota.

The industry comprises over 50,000 separate holdings, mostly grouped in areas of favourable soil and climate, adjacent to big consuming centres. As the Lea Valley is to London, so is the

Wirral to Liverpool, and the land behind Blackpool to Manchester and other Lancashire towns. The industry gives whole-time employment to more than 150,000 people, and has a host of part-time workers. Over half the whole-timers are proprietors, hard-working and often gifted growers, carrying on with the aid only of their wives and perhaps a son or daughter. It is they who will reap a bitter harvest of frustration and tragedy if Britain enters the Common Market with no protection for horticulture.

At the other end of the scale are a few giants, like the Rochford family of Broxbourne, Hertfordshire, London's tomato men for 80 years. Their heated glasshouses, dotted about the Lea Valley, cover 80 acres, and current replacement value is £18,000 an acre. The tragic meaning of "Common Market" to a fine family firm was given to me by Lieut.-Colonel Lionel Leach, 47, son-in-law of the Lea Valley's grand old man, Mr J P Rochford, now 79. "If the Government decides it is in the national interest to go into the Common Market it must accept the fact that the horticultural industry is doomed, unless it is granted exemption from conditions of entry, as at present laid down."

Why, I asked him, can English tomato growers not stand up to free competition from Holland, which has no significant climatic advantage?

The reasons, briefly, are:

1. Cheaper labour in Holland, where an elastic arrangement of working hours gives the Dutch grower a 15 per cent advantage.
2. An enormous injection of cheap capital, with which the Dutch have built in the past few years a vast area of up-to-date oil-fired glasshouses, fed by cheap oil and equipped with automatic irrigation.

In the Lea Valley and other English glasshouse concentrations most of the houses are old and out of date. The tide of urban development has flowed out and engulfed them. But, because of Green Belt restrictions, it is impossible for owners to sell their land for development, and move out into new areas. If they could, many are ready to start afresh with modern equipment and fight the importers.

Budget hint alarms farmers

John Winter 26.7.61

Farmers were seriously alarmed last night at the Chancellor's hint that their guarantees may be pruned at the next price review in February. He said: "We shall have to look critically at the level of agricultural support during the 1962 review."

The level of subsidies on some of farmers' products is running, through no fault of their own, at unusually high levels at present. Their only comfort is that the Chancellor is bound, by pledges at the last election and recent assurances by Mr Christopher Soames, Minister of Agriculture, not to start dismantling the system of Exchequer support during the lifetime of this Government. These include maximum limits by which the guarantees can be reduced in any year.

The Chancellor is further limited, in the case of pigs and fat cattle, by Mr Soames's promise that increased prices last February would be maintained for two years. Although the total value of the guarantees is £1,279 million, the actual bill for subsidies paid out was expected, at the last review, to be around £280 million this year.

Sharp falls in prices, which farmers have since suffered, may raise it well above this figure. The two products which have caused greatest anxiety are barley and beef. Barley has been depressed through huge quantities dumped at cut-throat prices in Britain, chiefly by Russia and France. Beef prices have been "on the floor" since an increased guarantee of 10s a cwt in February, and the early flush of grass this year released a flood of fat cattle on to the markets.

Soames's meat bill shocks Commons
John Winter 15.12.61

An increase of £78 million on the Government's estimate for farm subsidies shocked MPs yesterday. The major culprit is meat, with a jump of £67 million. The figures exceeded even the gloomiest forecasts by the farmers themselves.

Agriculture Minister Mr Christopher Soames came in for a brisk roasting from Labour members when he tried to explain in the Commons how the meat bill had soared. The new estimates will bring the year's bill for agricultural support to about £350 million, easily the highest since rationing ended. Fat cattle took £35 million of the excess, sheep £13.6 million and pigs nearly £18 million, and the staggering build-up, he said, followed an unprecedented chain of circumstances.

First, there was a big increase in the flow of cattle – the estimate for the year is 2.5 million against last year's

2 million – and prices slid from 141s a cwt in April to 127s 6d in May, 109s 9d in June, and 100s in July. By this time the taxpayer was finding 60s a cwt, or about £30 an animal to make up the farmers' price. Just as the price began to rise in August a bumper crop of lambs, 800,000 more than last year, knocked the bottom out of the market again.

The final blow was heavy imports of pigs; which caused the bacon market to collapse in September.

Cattle will cost £50,800,000, sheep £32,700,000 and pigs £40,200,000.

Challenged by Mr T F Peart (Labour, Workington) that the Government's support system had broken down, because both farmer and consumer had suffered, Mr Soames denied the butchers were the only gainers. Of the £76 million estimated increase over last year's fat-stock payments, about £31 million would go to farmers, because they produced more animals and had had higher guarantees for cattle and pigs, said Mr Soames.

Consumers were expected to save £35 million on cheaper meat and bacon. That left £10 million for the distributors, who had handled considerably more home-killed meat. But the cost of living index showed that from August to October housewives paid 5½ per cent less and the fall for the whole year would be only 4 per cent, while farmers' market prices were 15 per cent down.

Mr Soames warned farmers: "It is a matter for serious concern which we shall be bound to take fully into account at next year's price review."

CROPS and MACHINERY

The vital statistics of dolly the potato
John Winter 14.11.59

Housewives who do not like the vital statistics of some of the potatoes they see in the shops are disappointing the Potato Marketing Board. Good crops this year have brought prices down, and the Board expected a boom. Instead the demand has been sluggish.

Why? Because the ladies just do not fancy the shape of some of the potatoes. These potatoes – known to the trade as dollies – have a mis-shapen appearance because of the summer drought or because they have been affected by scab.

But despite this they may be perfectly sound inside. "It certainly seems that housewives are more concerned with the appearance than price," said Mr J T Newton, secretary of the National Federation of Fruit and Potato Trades. "Wholesale potato prices are so low they are scarcely economic," he added.

The steady rise in demand for pre-packed potatoes, which are washed and selected, but cost more, emphasises the housewife's preference. More than 12 per cent of all potatoes are now sold pre-packed, against 10 per cent in January.

Potato men mop up
Daily Mail Reporter 4.6.60

Another big load of unwanted potatoes is to be taken off farmers' hands. They are the remnants of last year's crop, made almost unsaleable by heavy imports of new potatoes. Probably 50,000 tons will be bought by the Potato Marketing Board at £9 to £15 10s per ton.

"It is a mopping-up operation," Mr Roscoe Herbert, general manager of the Board, said yesterday. The Exchequer will foot the bill jointly with the Board.

Worst-ever season for growers
John Winter 2.12.60

An inquiry into the plight of Britain's horticultural industry has been ordered by the National Farmers' Union. The inquiry follows the worst season market gardeners have ever experienced. With a record crop of 300,000 tons, dessert apples are at their cheapest since the war.

Yet producers' costs in the last ten years have leapt by more than half. Examples of rising overheads which are crippling the country's 40,000 market gardeners were given at a Press conference yesterday by Mr Eric Gardener, chairman of the NFU horticultural committee. Wages were up from 108s a week in 1951 to 160s today, with a further increase due in January. Fuel prices for greenhouse heating were up from £5 to £8 a ton, he said.

The farms spring to life again
John Winter 20.2.61

Thousands of farm tractors will roar into action this week in the biggest mass cultivation offensive for years. In large areas rain has kept the ground unworkable since last summer, and farmers have only a few weeks to prepare millions of acres for corn crops. High temperatures and sunshine last week brought a lot of land into condition for ploughing.

As the tractors roar in, fertiliser and seed firms will start a delivery blitz.

Shuttle services of road and rail trucks will run from their warehouses to the country merchants. Orders are flooding in and farmers are demanding supplies at once. Because so much nutriment value was washed out of the land by rain and floods, they will have to use extra fertiliser. One leading firm is advising an extra 1cwt of nitrogen to the acre for corn crops.

The long hold-up has seriously upset the balance of cereal varieties, and winter wheat sowing was 900,000 acres below average. Although farmers will fill some of the gap with spring varieties, a big increase in barley, more reliable for spring sowing, is inevitable.

The NFU gave these reports of the ploughing offensive yesterday:

• *South-east:* Farmers are working flat out on lighter land, which dried first. Special trains are carrying chemical fertilisers to strategic unloading points.
• *South-west:* Higher land is workable, but heavy soil still needs more drying. Some farmers are lifting last year's potatoes.
• *Cornwall:* Spring broccoli is maturing too early, and road convoys are rushing it to Bristol, Birmingham and Manchester.
• *Wales:* Farmers are calling the mild spell a "phoney" spring. Most of their ground needs another week of dry weather to make it workable.
• *East Midlands:* Farmers with land which has been waterlogged for months are ploughing it and taking a chance on frost to break up the clods.

Poison fields wipe out birds

John Winter 23.2.61

Hundreds of thousands of wild birds will die before spring is out – poisoned by eating seeds treated with toxic chemicals. The death rate will be greater this year than ever before, because the bad weather hold-up in winter farm work means more spring sowing than usual. Of different "kills" investigated last year some were caused by seed dressings, but most by chemical sprays. Wood pigeons, pheasants, rooks and crows, chaffinches, partridges and 27 other varieties of birds were affected.

The figures were given in London yesterday by the joint committee of the Royal Society for the Protection of Birds and the British Trust for Ornithology. Chemical firms were asked to intensify their search for chemicals which were more selective, and to label toxic substances dangerous to wild life.

Mr Peter Conder, secretary of the committee, said casualties may be expected among dogs, cats, foxes, squirrels and predatory birds, which feed on poisoned birds. There was also a danger to humans. "Some birds, especially pigeons, may have received a sub-lethal dose of toxic seed dressing. If they are subsequently shot and eaten, the results can hardly be pleasant."

Mild spell sets potato puzzle for farmers

John Winter 24.2.61

Spring in February, following months of incessant rain, has brought a freak clash of two seasons on potato farms. In Lincolnshire, the leading potato-growing county, farmers were yesterday harvesting last year's main crop, while their neighbours sowed early varieties for this year's harvest.

The mild spell has also upset forecasts of a big surplus from the record 1960 crop. High temperatures are causing rapid deterioration among potatoes in clamps, and if they continue, many which were being kept for sale in April and May will be unusable.

Absence of frost, on the other hand, has kept unharvested crops sound in the ground. Farmers now digging them are surprised at the weight of usable tubers coming up.

"The situation is quite chaotic," said Mr John Piccaver, of Gedney Marsh, near Spalding, a leading grower and member of the Potato Marketing Board. "Potatoes dug and clamped in wet condition are in a bad way. Blight is spreading through them. Some crops that are being dug four months late, however, look excellent, and are yielding up to six tons of good potatoes to the acre."

The Potato Board says it is impossible to forecast whether supplies of good potatoes will last out. Prospects of a 750,000-ton surplus, feared in November, have vanished. It is now touch and go whether there will be any to spare.

Poison will still be sprayed on your food   *John Winter 24.2.61*

Poisons will continue to be sprayed on crops, despite the ban on alkali arsenites announced by Mr John Hare, Minister of Agriculture. Without them millions of people would starve through shortage of food, it was claimed yesterday by the Association of British Manufacturers of Agricultural Chemicals.

But the Association has agreed voluntarily to the ban on arsenites, and members have already stopped their manufacture.

"The civilised world must learn to live with agricultural chemicals, or revert to a food-rationing system far more rigid than was ever experienced during the war," states the Association. "The ration would be in the form of weevil-infested flour, scabby and maggoty fruit, and disease and pest-ridden vegetables. These chemicals are the tools of the progressive farmer. If insects were allowed to develop unchecked, they could master the world, for they transmit dangerous diseases."

The statement was issued, said Mr George Huckle, chairman of the Association, because there had been so much ill-informed criticism of the use of toxic chemicals. "We felt this had reached a stage when, if we did not answer it, people would think we had no answer. We wish to put the matter in its right perspective."

Mr Huckle said there were good reasons for the decision announced by Mr Hare that arsenite spraying of potato foliage would be permitted on this year's crop, but not afterwards. Large stocks of arsenites had been manufactured, and it was difficult to provide substitutes. "Use of existing stocks is not solely a commercial consideration. The disposal of arsenical substances is extremely difficult, because there can be no question of dumping them," he said.

Alternative substances should be available in quantity next year, but they looked like costing about three times as much as sodium arsenite. Sulphuric acid, widely used before arsenites, cost about twice as much, and was unpopular because of its corrosive nature.

Potatoes fail 'new' test
John Winter 11.4.61

A test of "new" potatoes showed they were lifted in November and came from Cyprus. It was made by the Horncastle branch of the Lincolnshire NFU. Now the county headquarters, representing

the growers of a fifth of Britain's potatoes – is considering action to regulate
imports.

Good home-grown potatoes are being
ignored in favour of the so-called new
ones, it says, yet the quality is inferior,
the texture poor, they cannot be scraped,
and the taste is not so good.

Mirror-eyed bird scarer soars like a hawk
John Winter 28.4.61

A bird scarer in the form of a dummy
hawk, attached to a hydrogen-filled
balloon, has been invented by Mr Arthur
Knights, of North Walsham, Norfolk.

The "hawk", made of lightweight
material and with small mirrors for eyes,
can be flown at various heights up to
about 50ft. It sways in the wind and is
claimed to keep other birds away from
crops.

Top-level talks on plight of barley market
John Winter 15.6.61

Talks on the barley market were held
yesterday between Mr Christopher
Soames, Agricultural Minister, and
Mr Harold Woolley, president of the
National Farmers' Union. A statement
says there was full discussion of the
causes of the present difficulties and
the way to strengthen and stabilise the
market in the future.

Mr Woolley told the Minister that the
NFU intended to make an immediate
application under the Customs Duties
(Dumping and Subsidies) Act. The
Minister promised that the Government
would give it immediate consideration.
In the case of suspected dumping of
goods in Britain, the Board of Trade
undertakes to find out if material injury
to our own markets is threatened. In
some cases goods are dumped below the

market price to get rid of them. In other
cases a subsidy is concealed.

This year's harvest will be smaller
John Winter 20.7.61

The harvest outlook was given by Earl
Waldegrave, Parliamentary Secretary to
the Ministry of Agriculture, in the House
of Lords yesterday.

Winter wheat is about 900,000 acres
down: total 510,000 acres [*sic*]. Most is
likely to be made up by spring sowings.

Barley forecast in March was up by
259,000 acres to 3,318,000.

Oats are down to 928,000 acres, a
drop of 160,000.

Cereals look promising but some
spring-sown crops are making poor
growth.

Hay yields should be a little higher.

Potato sowings are down by 80,000
acres and the yield per acre of early
crops is expected to be slightly below
average.

More join the barley revolt
John Winter 24.8.61

The barley price war hotted up yesterday, as an avalanche of grain poured
from combines in thousands of harvest
fields. More farmers joined the revolt
against selling at cut prices, and
Britain's second-biggest farmer co-
operative offered its 5,000 barley-
growing members a firm forward-buying
contract, with cash in advance of
delivery.

With forecasts of a record 4,500,000-ton
crop, offers have dropped to 16s a cwt –
2s 3d below the price indicated by the
barley working party of the NFU, corn
merchants and feed compounders. Unless
farmers halt the slide, the NFU fears the
subsidy bill will be even higher than last

season's record £34 million.

The offer came from Southern Counties Agricultural Trading Society, with a £9 million turnover and 13,000 total membership, and territory from Wiltshire to Kent, covering some of the biggest barley areas. Members who keep barley off the glutted market for delivery when the society wants it will be paid a price in line with the working party's figure. To meet farmers' need for cash, they will receive 90 per cent of the price 28 days after signing the contract, and those with inadequate farm storage can use the society's two silos with 10,000 tons capacity.

"We shall encourage farmers to hold barley until the end of the year, when the guaranteed price rises," said Mr Eric Haken, managing director. "The forward contract has been made possible by the Government's action in limiting barley imports and fixing a minimum price. We believe this is wanted by both farmers and the Government, and will also help feed compounders, who use the barley, and are contracting with us for planned deliveries."

To discourage a flood of grain at harvest time, 9d a cwt is deducted from the 27s 7d guarantee up to October. In November and December farmers receive the flat rate; in January and February, 1s premium, and from March to June, 1s 6d premium.

Britain to try out new-look Mercedes tractor   *John Winter 29.8.61*

British farmers will soon be able to drive a Mercedes-Benz, even if it is only for ploughing. The first Mercedes-Benz tractor seen here will take the field next month. It is the Unimog, an all-purpose model, and will be demonstrated on Lord Montague's estate at Beaulieu, Hampshire. Revolutionary features – to English eyes – are four wheels of equal size, and a driving cab to seat two.

Barley war farmers go on strike   *John Winter 30.8.61*

A leader of the National Farmers' Union broke all tradition in an ancient corn exchange yesterday and made a speech. At the same time farmers stopped business. The farmers' revolt against cut prices for barley had taken a new turn at Hitchin, Hertfordshire. The farmers refused to sell their grain at less than the 18s 3d a cwt official guide price.

Mr David Carter, county delegate to the NFU National Council and leader of the resistance movement, was applauded by more than 100 farmers and merchants after his speech. In it he said: "I believe the loyalty of both NFU and Corn Merchant Association members, with bonds of friendship founded on years of straight trading, will win the day."

Significantly, representatives of the national compounding firms, who have been blamed for low prices, were absent. They claim they have bought all their immediate needs already and can only accept additional supplies at lower prices, to cover extra handling and storage.

The revolt is expected to be copied – and it comes as the harvest crisis reaches its peak. As the pressure to sell mounts, the NFU is striving desperately to complete emergency plans – perhaps to hold barley off the market – to strengthen farmers' resistance and save the Exchequer from a monumental subsidy bill which, in turn, might react again to next year's guaranteed prices.

Bills win the harvest battle

John Winter 4.9.61

In the baking heat of the last week in August, the barley "war" was won and lost at Fairfield. Working ten hours a day amid the clatter and dust, Farmer Dring completed his harvest three weeks earlier than usual.

But in selling his barley, which made up nearly two-thirds of it, he was beaten from the start. He could not get the elusive "guide price" of 18s 3d a cwt, even for the first batch sold. But he considers himself pretty lucky to have been quoted 17s 5d for all the barley he intends to market straight from the field.

The rest of his barley is Pallas, the new atomic variety, which he grew on contract for a seed firm and will store on the farm until the firm is ready to collect it, probably early next year. Why did he not store all the grain, to take advantage of the guaranteed price premium offered for barley which is held by the farmer until January, or later?

"Chiefly because I want cash now to pay bills," he told me. "We always get an invasion of creditors and salesmen at harvest time.

"Secondly, the inducement in the guaranteed price is not sufficient to cover costs of storage and other losses now the Chancellor has raised interest rates. I don't quarrel with my merchant. He does not want barley at the moment because the compounders, to whom he sells, have stocked up for their immediate needs. But I do feel the compounders are exploiting the harvest flood, and their claim that what they save by buying cheap barley is returned to us in the form of cheaper feeding stuffs is just bunkum."

One move by Mr Dring to reduce his harvesting costs is the purchase of a grain dryer. Last year a contractor charged him three guineas a ton to dry barley down to the 16 per cent moisture level demanded by the trade. His old electric dryer handled only half a ton of grain an hour, with a man standing by. The new one, with a diesel-driven fan which blows air through a stack of corn sacks, will dry 20 tons a day, unattended. Fuel cost in good drying weather is 2s a ton. But it is a fairly costly investment – £650 – which Mr Dring has paid out of a capital account.

All the corn crops yielded surprisingly well, except 11 acres of spring wheat, which was poor. Winter wheat was nearly two tons an acre, and both winter and spring oats the same. The best field of atomic barley produced two tons to the acre, but on poorer land it only equalled the older Rika variety with 32cwt.

Keep barley off the market, say merchants

John Winter 26.9.61

Corn merchants, struggling to handle the biggest home barley crop ever grown, yesterday called on the Government to arrange for extra short-term credit to help farmers hold grain off the market at harvest time. This, said Mr Henry Nicholls, president of the Corn and Agricultural Merchants' Association, would help to reduce deficiency payments due to farmers by the Exchequer under the farm price guarantee.

The merchants estimate the 1961 barley crop at 4,750,000 tons, 12 per cent more than last year's record. Wheat, at 2,440,000 tons, is 550,000 tons down. But in spite of last winter's appalling sowing conditions, the total cereal harvest should be over 9 million tons, only 250,000 tons less than last year.

Barley, said Mr Nicholls, was the

problem crop. And for the first time it will supply practically all requirements for human consumption and animal feeding stuffs.

Two companies quit seed combine after trade row

John Winter 20.10.61

The giant National Farmers' Union seed firm, launched only eight weeks ago to buy and sell grass seeds to farmers, has lost two of its four partners before it has opened for business. The NFU revealed yesterday that two of the three old-established seed firms which joined with the union to form NFU Seeds Industries Ltd, a £250,000 company, have withdrawn from the merger. Their resignation follows bitter criticism from the seed trade.

The retiring firms last night denied the rift was caused by the attitude of fellow traders, but a spokesman of one said: "I think our withdrawal will go a long way to meeting objections of the trade." The two firms, both retailers, are Dunns of Salisbury, and Townsends of Stroud.

Remaining in the company is the wholesaling firm Hurst, Gunson, Cooper,

Tabor of Kelvedon, Essex. It will now have two directors on a board of seven, with the other five, including the chairman, appointed by the NFU. Mr Harold Woolley, NFU president, said yesterday the original plan for the NFU to provide half the capital would be retained, the other half coming from the remaining firm instead of the three.

Two new tractors to make the job easier

John Winter 28.11.61

Two new tractors designed to make the driver's job easier and capture export orders will be shown by Nuffield at next week's Royal Smithfield Show.

Both the "460" four-cylinder model (£650–£807) and the "342" three-cylinder (£595–£752) have automatic depth control. The driver can set his plough and other implements by lever, without leaving his seat. Another lever enables him to select the hydraulic lift independently of the power take-off, or operate both together.

The firm plans to treble output with the two models when its tractor production moves to the new BMC factory at Bathgate, West Lothian, next year.

 # LIVESTOCK and POULTRY

90 pig producers join in protest march

John Winter 13.11.59

Ninety angry pig farmers marched on Westminster yesterday carrying a 30ft banner, "British pig farmers demand justice".

The marchers, members of the British Association of Pig Producers, left their farms to travel from all parts of England on fog-bound roads and railways. Their

mission was to press for immediate Government action to save the English bacon industry. To the 14 MPs of all parties who received them they outlined the plight of their industry, which held a third share of the home bacon market only two months ago, and today holds little more than a fifth.

They came away with promises of support for the minimum action they consider necessary – if the English break-fast table is not to be laid wide open to imports when the first half of the 10 per cent tariff on Danish bacon is removed next July. This is an assurance *now* that the Government subsidy for pigs will be split at the February Price Review into separate guarantees for bacon pigs and those used for pork.

Members of both Conservative and Socialist party agricultural groups prom-ised help. Four members of the pigs committee of the Conservative group said they would bring the crisis before their committee immediately.

Mr Douglas Nicholson, association secretary, said: "Throughput of English bacon, which stood at 2,800 tons a week at the end of October, fell to 2,600 tons last week, and is estimated this week at 2,150. Multiply those figures by 14 pigs to the ton, and you see how pigs are dis-appearing by thousands a week from the bacon factories into the fresh pork and manufacturing trades. It is useless for the Government to remain silent until the Price Review. By February there will be no English bacon left to save."

The crisis has arisen from an overall shortage of pigs, coupled with a strong demand for pork. This has pushed the pork market price to an uneconomic level for bacon curers, and cut the guar-antee, spread over all pigs, to a few coppers a score. Confidence of bacon pig producers is further undermined by the shadow of the Danish tariff removal.

New farm ideas may mean better meat in the shops

John Winter 18.11.59

With a strip of sticking plaster across her forehead, herdswoman Molly Johnston today led out ten very important bulls into a straw-covered yard. The plaster concealed a two-inch gash which one of them gave her while she was scrubbing him last week at the British Oil and Cake Mills experimental farm at Barlby, near Selby.

"He didn't mean it," said Molly.

The audience of 80, who sat on piers of straw bales to watch her parade the bulls, included many of Britain's leading cattle breeders and livestock scientists. They came to see the bulls, all Beef Shorthorns from Scotland, start a 168-day test to determine their ability to produce calves from dairy cows which will give the housewife leaner and more tender steaks and sirloins.

Molly, who has worked at the farm for four years, is in charge of the bulls, many of which had never been on a halter before they arrived at Barlby four weeks ago. She feeds them with measured rations on a rigid timetable at 6.30 am, 10 am, 1.30 pm and 4.30 pm, and grooms and exercises them daily.

"We have found girls have a better temperament than men for rearing calves," said farm manager Pat Bichan. "When the same girl is managing them all the time they get to know her and behave better. Sometimes we have to change bulls over to herdsmen when they get older and need a firmer hand."

But the test bulls will probably remain under Molly's care until they leave next year to start breeding on farms and arti-ficial insemination stations.

The turkey factory: down at Middle Wallop they're already planning for Christmas 1960

John Winter 19.12.59

Your Christmas dinner strutted past me yesterday. It was looking fine. I was standing outside the white-walled office of a farm in Hampshire. The air was filled with the treble trilling of hen turkeys. The birds, in wire crates, were on the first stage of their last journey. Inside the office the telephone was buzzing. The eyebrows of the gumbooted farm manager shot up as he lifted it and listened.

"How many?" he gasped. Pause.

"Well, we'll do our best. But if you want any more we shall have to process the farm dog and send him!"

At the other end was the poultry buyer of a famous firm of multiple food shops. He had a contract for more than 90,000 turkeys from the farm, all quick-frozen, pre-packed, oven-ready. He had stipulated that the last of his Christmas dinners must be in his London cold store by December 14 – but that was before the avalanche of orders from eager housewives.

This year will see a record total of nearly 4 million turkeys on the Christmas market. If they are all sold it means that, for the first time in history, approximately half the population of Britain will sit down to a turkey dinner on Christmas Day. Well, here's one firm, and one of the biggest in the business, who have underestimated the demand.

"He's on the blower every day," said the farm manager.

The farm at Middle Wallop is appropriately named Brewery House, and at one time the local wallop was actually brewed there. But now it is owned by two brothers, Malcolm and Pat Bradley, who could claim only knife and fork contact with turkeys until they were demobbed from the Army and Navy respectively after the war. They are the biggest producers in Britain. They rear the turkey through its life, from the egg to the rock-hard, oven-ready table treat, vacuum-wrapped and packed in its neat carry-home carton. This year they hatched 250,000 poults, reared 100,000 themselves, and sold the rest to other rearers.

If they cannot meet the last-minute repeats of the multiple firm, who can blame them? When they started the contract, only two years ago, it was for 30,000 turkeys. Last year it rose to 50,000, this year to nearly twice as many.

Next year? Malcolm and Pat Bradley are doing their best. A vast new brooder house to hold 10,000 extra poults from three to nine weeks of age is already stretching its skeleton across a field. Their 5,000 breeding hens are already busily producing. Early chicks are twittering in the brooders. First batch was hatched in November. It is all done to a streamlined, labour-saving efficiency plan-pattern of the farming of the future.

Forgive me if I talk about next year's turkeys before you have eaten this year's, but there's an exciting prospect looming which could make current production figures look silly. There is hardly a limit to the type and size of turkeys this farm-factory can supply – anything from a neat little four- to five-pounder for two to a 40lb cock turkey for the grand hotel.

Part of their contract this year has been a trial run of 10,000 4½–6lb roasters . . . meaty little turkeys killed at 13–14 weeks. As an all-the-year-round alternative to the weekend joint for the average family they are novel, tasty, and a welcome change. They sell at 20 to 25s – a first-class buy.

But there is a snag. The cost of poults,

9*s* apiece, puts this kind of turkey out of the reckoning at present for rearers who do not breed their own stock like the Bradleys. "There is a ready demand for the little roaster at any season providing the price is right," said Malcolm. "But 25*s* seems to be the limit the housewife will pay."

So, undoubtedly, it is value that can build turkey business and there is no limit to the energy the Bradleys will use to get housewives to think and take turkey month after month. The other day, with fellow members of the British Turkey Federation, they went into Salisbury and ceremonially carved a 13-pounder, and stripped it to the bone. They showed the audience that 70 per cent of the total weight was meat, 9 per cent giblets, and only 21 per cent bone.

Then they made a 22-pounder into turkey sandwiches.

"The bird would cost 4*s* 6*d* a lb to buy, and it filled the rounds of sandwiches cut from twelve 2lb loaves," said Pat. "How's that for value compared with red meat?"

Killer virus infects the farm
John Winter 22.1.60

Virus from a deadly type of African foot and mouth disease has leaked out from a research station and infected cattle on a farm one and a half miles away. The leak, from the Animal Virus Disease Research Institute at Pirbright, Surrey, was admitted yesterday by the Ministry of Agriculture.

A full investigation into the apparent failure of the £1 million institute's security arrangements has already started, said the Ministry. The institute was started in 1924. This is the first outbreak traced to it.

The 13 beef cattle and nine pigs on Perry Hill Farm, Worplesdon, where the disease broke out, were destroyed a few hours after the outbreak was confirmed on Monday.

British AI men seek bulls
John Winter 15.7.61

While Customs investigators are trying to unravel the smuggling from France of the "bull that never was", Ministry of Agriculture livestock chiefs will fly to France on Monday to select bulls which really are, for import into Britain.

A brains trust of cattle breeding will make a preliminary selection for the most controversial importation of cattle in farming history. They are looking for sixty Charollais bull calves, from which half will be finally chosen in autumn for purchase by the Ministry. After a strict quarantine the calves will be distributed among artificial insemination centres for breeding tests with English dairy cows, to find whether the resulting calves produce better beef than home-bred ones.

Leading the delegation is Mr W P Dodgson, Ministry chief livestock husbandry officer, and the party includes AI breeding experts from the Milk Marketing Board and private AI centres.

The Customs investigation follows the appearance at English shows this summer of a calf born to an Ayrshire cow on a Somerset farm last September. The sire was a Charollais, and the semen was smuggled to Britain when imports were banned, and a Government committee was inquiring whether a test should be made.

The Little Lion may be taken to the cleaners
John Winter 14.8.61

The Egg Marketing Board is trying to

persuade farmers to accept a rule which many housewives learned from their mothers – that an unwashed egg is more likely to remain good than one that has been washed. This is because moisture may penetrate the shell and turn the egg bad.

The board has tested eggs picked at random from packing stations all over Britain.

It was found that after one week bad eggs in each box of 360 which had been machine washed averaged 30. In the second week another 48 went bad. The figures for eggs which were naturally clean, or had been dry-cleaned, were under one in each week.

Now the board wants farmers to dry-clean their eggs when necessary, but ideally to try for eggs which are naturally clean.

Next summer it hopes to introduce a differential-payments system, which would give farmers a bonus – probably up to 3d a dozen – for unwashed eggs. This would be financed by a deduction in the price of washed eggs. A simple ultra-violet test would reveal evidence of washing, and spot checks would be made at packing stations.

Although all the farmers' unions support the move, there is strong opposition from Lancashire, the leading egg production county. Big producers, who put all their eggs through washing machines, say dry-cleaning would be uneconomic.

The Board points out that the major egg-exporting countries ban the sale overseas of washed eggs. In some areas housewives are already asking for unwashed eggs. Eventually they may be able to buy them in the shops at a slightly higher price.

Charollais get full security treatment
John Winter 1.11.61

A disease security curtain will be drawn so tightly round the Government's 30 Charollais bulls when they arrive from France next week that only authorised handlers will see their arrival. They will be flown in four batches into Northolt, but will not be allowed to put even a hoof on the tarmac.

They go straight from the aircraft into cattle wagons, and the official reception party will see them at the Ministry of Agriculture quarantine station at the Royal Albert Dock. After 28 days' quarantine the bulls will be distributed among AI stations throughout the country, and beef-breeding tests with English dairy cows will start in the New Year.

MP acts as cattle slaughter mounts
John Winter 17.11.60

The Government is to be asked to test a system for controlling foot-and-mouth disease which has been successfully used in France. Nearly 20,000 cattle, sheep and pigs have been slaughtered in this country in ten days to combat the disease.

Mr Mark Woodnutt, Tory MP for the Isle of Wight, will ask the Minister of Agriculture, Mr Christopher Soames, whether his department will investigate the system. "The French authorities are spreading on their pastures a granulated mineral substance which grows on marine coral off the Brittany coast," said Mr Woodnutt yesterday. "The system has been under test in Brittany for six years, and a report to the French Academy of Agriculture states that it does build up an immunity in cattle."

When Mr Woodnutt submitted a report

to the Ministry in April he was told the methods of control in France differed from those in this country, where foot-and-mouth disease is not endemic. A new outbreak in Norfolk yesterday brought the total since the epidemic started to 114, involving the slaughter of 7,500 cattle, 7,100 sheep, and 4,250 pigs.

Commons probe on Charollais plan
John Winter 14.12.61

Ten questions are lined up for Mr Christopher Soames, Minister of Agriculture, in the House of Commons today – asking what he intends to do about the French bulls his livestock boffins brought over in October.

Last Friday, the 30 Charollais bulls bought for beef breeding tests with English dairy cows should have been sent to artificial insemination stations. Instead, three were found to have a blood infection and were shot. The others have been given three weeks' extended quarantine, which may be extended to six months to meet breeders' fears.

Four questions for Mr Soames speak of a disease "outbreak" at the quarantine station. One calls bluntly for the bulls' destruction. Others that they should be sent back to France. The facts are that no outbreak occurred, and there is no danger that the disease can affect our own cattle. The Charollais, if released, will spend their lives in AI stations, and have no contact with the cows with which they will be tested. The disease in question has not been found to be transmissible in semen.

 # EMPLOYMENT

More pay
John Winter 23.3.61

New minimum wage rates for agricultural apprentices and craftsmen were fixed by the Agricultural Wages Board yesterday. Apprentices will get from £2 19s 6d at 15 to £7 11s at 20.

Too many small farms
John Winter 11.4.61

A quarter of Britain's food is being produced by people "little more than peasants, sweating out each day without thoughts or hope of tomorrow," Mr Travers Legge, 1959 Fisons Award winner, told the Country Landowner's Association at Cirencester, Gloucestershire.

"There are 166,000 farms of under 20 acres. If we could reach a stage at which the minimum was 35 to 40 acres of the better land, we could offer competition in the Common Market which no country in Europe could match."

The farmers who get less than farm labourers
John Winter 14.4.61

Hard-up hill farmers should have their incomes subsidised by the Government *up to the pay of farm labourers*, say two farm economists in a book published today.

They made a survey which showed that one-quarter of the hill farmers in Wales had an income of £250 or less, which had to cover their manual labour and that of their wives, interest on working capital, and a reward for management. The survey was four years ago, but farm incomes have risen only 10 per cent since. Most of the 200 farms had an annual production below £2,000 – £500 less than the minimum which the joint authors, E A Attwood and H G Evans, think necessary to cover labour, interest and management.

They say in *The Economics of Hill Farming* that only 42 per cent of the smaller farmers were willing to take on extra land. This would be the obvious solution on farms too small to produce the minimum output.

The authors, who are on the staff of the University of Wales, have a remedy: money already paid by the Government in improvement grants should be channelled from bigger farmers to the little ones. Income subsidies to raise the farmer up to farm workers' rates should be limited to his lifetime. Production on farms capable of improvement should be boosted by grants of 80 per cent, or even higher, of the capital cost of improvements. Farmers on bigger and more economic holdings would have their grants scaled down.

Wanted! 200 teachers for farm colleges
John Winter 15.4.61

Wanted urgently: two hundred extra teachers, with very special qualifications, to teach the rising number of students who hope to make a career in agriculture. Sir David Eccles, Minister of Education, yesterday announced steps to complete a comprehensive education system in farm institutes and technical colleges, following the report of an advisory sub-committee. Both he and the Minister of Agriculture, Mr Christopher Soames, urge local authorities to take early action.

Teaching strength should be raised from 550 to 750 in the next three years, says the report. This means doubling present recruitment. Teachers will require sound technical and scientific knowledge plus practical farming experience. Other proposals are: separate courses both whole- and part-time for women; and specialist courses for students who wish to take up horticulture, poultry or forestry.

Welcoming the plans, a National Farmers' Union spokesman said last night: "The increasing level of skills demanded by modern farming in all its branches makes it imperative that the smaller labour force shall be highly trained."

Apprentices get a new deal
John Winter 2.5.61

Farm apprentices and the farmers who train them get a new deal today under an apprenticeship scheme backed by their unions. Boys and girls will get about 17*s* a week *below* the standard age rate during their three years' training. But on passing out at the end they will become entitled to 186*s* a week – 17*s more* than an ordinary farm worker.

The idea is to increase the incentive for country boys to take up farming and recompense employers who train them and allow them time off with pay to go to technical classes.

More and more training opportunities
John Winter 29.6.61

Record number of evening courses on such varied subjects as tractor maintenance and handling are being run by Norfolk Young Farmers' clubs.

John Winter 11.7.61

Improvements at the Institute of Agriculture, Moulton, Northamptonshire, will cost £260,000. They will include a 350-acre stud farm.

John Winter 3.8.61

150 students – half the number of young people going into farming each year in Nottinghamshire – will be trained at the county's farm institute at Brackenhurst when a £113,000 extension scheme is finished.

The farm workers have to wait

John Winter 17.10.61

Five hundred and fifty thousand farm workers will have to wait until November 10 before being told whether they will get higher wages and shorter hours. A meeting of the employers' side of the Agricultural Wages Board decided in London yesterday to "make no offer meantime".

The two sides will meet again on November 10 when it is hoped a decision will be reached, said a Ministry of Agriculture spokesman. The current minimum for men is £8 9*s* for a 46-hour week. Women aged 18–20 are entitled to a minimum rate of £5 18*s* 6*d* in most counties in England and Wales.

Farm rise defies pay pause

John Winter 11.11.61

An increase of 6*s* a week was awarded to farm workers last night in defiance of the Chancellor's pay pause. But the increase will be deferred until February 26, six weeks later than the earliest date at which it could have been paid.

A claim for a shortening of the 46-hour week was rejected.

The rise – granted by the Agricultural Wages Board – brings the minimum rate for men to £8 15*s* a week. Proportionate increases are awarded to women and juniors. A total of 544,800 workers in England and Wales will benefit.

Said a Treasury official: "The Government would like it clearly understood that its attitude to the wage pause is in no way affected by the action of this independent board, which has carried out the functions entrusted to it by statute." The board is completely autonomous. Unlike the 60 wages councils in industry, its decisions cannot be challenged, and the Minister of Agriculture has no power to vary an award, or delay its operation.

Mr Harold Collison, secretary to the Agricultural Workers' Union, said: "We are bitterly disappointed. We consider the award is completely inadequate, and see no reason why farm workers should not have a shorter working week."

Workers get a new pension scheme

John Winter 22.12.61

Farm workers can boost their State pensions by at least £175 a year by contributing 2*s* 6*d* a week from the age of 20 to a new voluntary pension scheme announced yesterday by the National Farmers' Union. The scheme, starting in January, is backed by the National Union of Agricultural Workers and is based on equal contributions by workers and employers.

Workers can increase their entitlement by extra payments, in multiples of 2*s* 6*d* a week, for pensions payable at the age of 65 for men and 60 for women. Pensions are transferable when a worker changes his job, if his new employer is in the scheme; if not, his pension rights are preserved. The workers have three representatives on the nine-man pensions scheme board.

PEOPLE in the NEWS

The potato rebel will be tried by his colleagues   *John Winter 8.1.60*

Farmer Mr Jack Merricks, rebel member of the Potato Marketing Board, yesterday received a summons to appear before the disciplinary committee of his own board. He is charged with failing to make a return of the area of potatoes he grew in 1958, and faces a penalty of up to £200. He will be "tried" later this month by six of his fellow board members.

Although, as the champion of free marketing, he was in constant trouble with the board after its reconstitution in 1955, this is the first summons since his surprise election to the board in October 1958. "It is rather sad, isn't it?" said 49-year-old Mr Merricks at his farmhouse facing the Channel at Winchelsea, Sussex, yesterday.

Since his election, on a pledge to potato growers to work for the board's abolition, Mr Merricks has done his best to avoid trouble. Last year, although opposed to the principle as strongly as ever, he made a return of his potatoes and had it certified by a chartered surveyor.

"I knew that, under a special rule, all board members were to have their potatoes measured, to ensure they got the same treatment as everyone else," he said. "But I sent a letter with my return, suggesting that, because it was certified, it would be unnecessary for the board to put growers to the expense of sending a man to check my acres. However, he came, and spent three days going behind every heap, stack and ditch, looking for potatoes."

Mr Merricks, who has already received a bill for a £363 levy for growing 33 more acres of potatoes than the board says he should, intends to defend himself before the disciplinary committee.

Victory for the fighting farmer
John Winter 27.9.60

For 25 years Mr John Salter Chalker has fought to clear British farms of deadly bovine tuberculosis. As a farmer he deplores the waste of 24,000 cattle slaughtered every year. As a committee member at a big sanatorium at Peppard, near Reading, he was even more appalled at the sight of children crippled through drinking infected milk.

In 1936 Mr Salter Chalker's herd of Guernsey cattle at Twyford, Berkshire, was one of the first in Britain to be certified TB free. Yesterday that herd was sold, because its owner is cutting down his farming. But he was not at the sale. Instead, he stood in a Whitehall conference room at an informal party thrown by the Minister of Agriculture, Mr Christopher Soames, to celebrate victory in the fight against bovine TB.

On Saturday the whole of Britain becomes attested. The area-by-area eradication scheme, compulsory since 1951, has cost £130 million. Two thousand five hundred Ministry and private veterinary officers have tested every herd in the country. They have ordered the slaughter of 110,000 cattle, for which farmers have received £6,500,000 compensation. More than 95 per cent of farmers joined the testing scheme voluntarily.

White-haired Mr Salter Chalker was the only man at the Minister's party who had been in the fight from the start. "I express the gratitude of the Government and the whole country at this great achievement of ridding ourselves of bovine tuberculosis, which is due to the excellent co-operation of farmers and the veterinary profession," said Mr Soames.

He singled out Mr Harold Woolley, president of the National Farmers' Union and Mr Salter Chalker, chairman of the union's animal diseases committee. "This is the culmination of the most important part of my life's work," said Mr Salter Chalker, who owns four other farms with two herds of British Friesian cattle. In one of them 20 animals were killed while Ministry scientists perfected their testing technique.

"I never counted the losses," he told me. "I was so shocked to see the children suffer through infected milk. I determined that farmers must rid their herds of this terrible disease."

Girl advises farmers

John Winter 8.2.61

Farmers were told yesterday that housewives are fed up with buying potatoes which are damaged, dirty, sub-standard, and mixed in size.

Twenty-five-year-old Miss Susan Marsden-Smedley, barrister and member of a Derbyshire farming family, told the Power Farming Conference, Brighton: "Potatoes should be cleaned by dry-brushing, because washing spoils their keeping quality. If pre-packed they should be labelled with the exact weight, date of packing, name of variety, and whether suitable for mashing, boiling or roasting," she added.

Rebel Jack attacks one-crop system

John Winter 10.2.61

Rebel farmer Jack Merricks, who opposes marketing boards, yesterday attacked the one-crop system. Mr Merricks, twice winner of the first prize for the best farm in Kent, said the system had produced the dustbowls of America and eelworm infestation of potato land in the English fens. Similar conditions were developing through over-cropping with sugar beet.

"What we want is more variation in cropping," he added. "If you use chemical weed-killers, you may manage to grow a greater succession of the same crop, but the land has a long memory. When you get your land in a mess it takes a long time to get it back again."

Mr Merricks was replying to agricultural service expert Michael Dodson at the National Power Farming Conference at Brighton. Mr Dodson said farmers would have to scrap old-fashioned theories about rotations and concentrate on continuous growing of profitable crops by using chemical weed-killers, which were ridiculously cheap.

Rebel Jack has a dig at the potato plan

John Winter 25.3.61

Rebel farmer Jack Merricks, 50-year-old champion of farming freedom, last night attacked the Potato Marketing Board, of which is a member, for its move to treble the levy on crops. The Board's proposals, which reached its 76,000 registered producers yesterday, were a bigger shock than they expected.

It wants the ordinary levy limit raised from £1 to £3 an acre and the charge for excess growing from £10 to £25 an acre.

Mr Merricks, whose home is at Icklesham, Sussex, said the plan would raise farmers' dues from £800,000 a year to

£2,500,000, make potato growing more restrictive and monopolistic, interfere with crop rotations, and increase the risk of eelworm infestation. The Board's operations, he said, have been so inefficient that the Government has had to come to the rescue financially and many farmers no longer believed they have a worthwhile price to guarantee.

Hilary will sleep beside her goats
John Winter 13.6.61

Hilary Bokenham, a 15-year-old schoolgirl, is to take three goats 250 miles from Cornwall to exhibit them at Tunbridge Wells and South-Eastern Counties Show on July 18. She will sleep alongside her goats at the showground and rise at dawn to prepare them for the ring.

Her school friends were so keen to show the goats that they went potato picking to raise money to pay the entrance fees. Hilary is a member of the Young Farmers' Club at St Gorran School, Gillan, near Helston.

Farmer Merricks prepares for war
John Winter 28.8.61

Farmer Jack Merricks faces his toughest battle this autumn in his war against agricultural marketing boards. He will be making his last stand against the Potato Marketing Board's plan to treble farmers' levies and at the same time be fighting to keep his seat on the board.

Final stages of the board's plan to gain Parliamentary approval for the collection of extra money needed for a £3 million-a-year market support fund are almost certain to clash with the triennial election of special members. At the last poll in 1958 Mr Merricks shocked the board by winning one of the three seats for England and Wales at the expense of Mr

Gwilym T Williams, then its chairman.

Election nominations close on September 18, and Mr Williams, 47, who farms at Newport, Shropshire, has been nominated by the National Farmers' Union, of which he is vice-president. "I feel I owe it to myself and the union to have another go," said Mr Williams. "But I have no ambition to regain the board chairmanship. My NFU responsibilities are substantial, and if elected I should be happy to serve under the present chairman."

Mr Merricks, 51, who has a 470-acre farm at Icklesham, Sussex, and farms another 2,000 acres with his brother, will have strong backing from farmers who do not want the board to be a union closed shop. "So many producers have urged me to stand again that I feel I must," he said. "Much of this support has sprung from my opposition to the increased 'tax' the board wants to impose on potato growing."

Tomorrow the board will learn whether the minimum of 1,000 farmers, growing 20,000 acres of potatoes among them, have demanded a poll on the financial amendments. If not, Mr Merricks will lodge objections with the Minister of Agriculture, who must either accept or lay them before a public inquiry. The amendments must then be approved by both Houses of Parliament.

Hat-trick for NFU in potato election
John Winter 26.10.61

The Potato Marketing Board election result, declared yesterday, is a triumph for the National Farmers' Union. Their team of three candidates won the three vacancies for special members for England and Wales in a 25 per cent poll, the highest recorded in these elections.

Rebel board member Mr Jack Merricks, of Icklesham, Sussex, lost his seat to Mr "Bill" Williams, NFU vice-president, who was chairman of the board when Mr Merricks defeated him in the corresponding poll three years ago.

The figures were: John Bennion, Stackpole, Pembrokeshire, 13,020; William Ruane, Ely, 12,726; Gwilym T Williams, Newport, Shropshire, 12,123; Jack Merricks, Icklesham, Sussex, 10,538; Wallace Day, Barnstaple, 4,370.

Retiring member Mr Wallace Day, who opposed the three official NFU candidates and came bottom of the poll, said the union had used "all possible means" to get their candidates elected to the special member seats. He sent a telegram to the Minister of Agriculture, Mr Christopher Soames, yesterday asking that the vacancies should not be declared filled until complaints in a letter dated October 14 from Mr Day had been investigated.

Farmer Merricks disowns mystery potato crop

John Winter 14.11.61

For an hour yesterday Farmer Jack Merricks listened while the disciplinary committee to the Potato Marketing Board, on which he lost his seat last month, investigated a charge that he failed to make a return of potatoes he grew this year.

Then the 50-year-old rebel farmer calmly revealed that the potatoes the six board members were discussing were not his. The 26 acres on which they were grown at Lodgelands Farm, Rucking, Kent, were let last December, he said.

Who then grew the potatoes, which board supervisor Albert E Morris had discovered by peeping over the hedge? Mr Merricks declined to enlighten the "court".

Mr Derek Syrett, solicitor for the board: "Why did you not inform the board that this land is not in your occupation?"

Farmer Merricks: "I have some farming to do, as well as form-filling."

Mr A G Wright, a board member: "Did you let the land for someone to grow potatoes?" "I don't know, Mr Wright. Why should I tell you about my private business? Do I ask you what you have done with your land?"

Farmer Merricks admitted that he had not complied with requests made in June and August to make a return for his own potatoes. He sent his return last Friday, saying the only potatoes he had grown were in the garden of his house at Icklesham, Sussex.

He was fined £5 for failing to make returns in time. It was the smallest of four penalties imposed on him since he first clashed with the board in 1956. But the committee imposed a £40 penalty on W Merricks and Sons, the family farming company in which Jack is a partner, for a similar offence. Fifteen other growers were fined a total of £152.

In the Commons yesterday Mr Christopher Soames, the Agricultural Minister, said that 8,512 objections had been made to the proposed alteration to the Potato Board's marketing scheme. So a public inquiry is to be held.

The board wants to treble the £1-an-acre levy on farmers to help finance a support fund to "prop up" the market in years of surplus.

GENERAL

The hooligans: armed gangs bring fear to farmers

John Winter 27.1.60

Armed gangs of trigger-happy hooligans are roaming the countryside, defying farmers who try to stop them and even threatening to attack them in their own farmyards.

This picture of gunlaw in the green heart of England and Wales was painted at the annual meeting of the National Farmers' Union in London yesterday, when a unanimous demand was made for legislation to stop it. Farmer after farmer from widely separated areas got up to recount incidents in which guns and knives had been flourished and sometimes used.

Mr W Howard Gibbon, 50, of Bedwas, Monmouthshire, former Rugby player and boxer, showed me a scar on his hand. A youth trying to steal roofing sheets gashed him with a sheath knife. "Although that case was attempted robbery, most of the examples reported to us are sheer bravado," he said. "But they are no less dangerous and worrying to farmers."

Mr H John Jones, 40, of Blackwood, Monmouthshire, said reasons for the trouble which had grown up in the past 18 months were: easy-to-get gun licences, the use of cars, restrictions which prevent police from entering land to deal with trespassers, and light sentences by the courts.

Mr Gibbon gave these examples of other farmers who have been threatened:

One was trying to eject a trespasser with a gun when it went off in the scuffle.

The trespasser was brought before the magistrates and fined. Shortly afterwards the farmer's buildings went up in flames, with the loss of some of his livestock. "On the cause of the fire, your guess is as good as mine," he said.

Another farmer, trying to put a man off his land, had a gun muzzle thrust at his stomach. "Now try to put me out," said the trespasser.

Farmers from Shropshire, Glamorgan, Yorkshire and Northamptonshire said they, too, had gun law in their counties.

No foxes, so hunts close

John Winter 23.2.60

Three East Anglian hunts have abandoned hunting for the season because of a shortage of healthy foxes. The Suffolk Fox Hounds decided to stop yesterday. Fitzwilliam and the Newmarket and Thurlow Hunts have already closed.

The premature end to the season has been caused by a mysterious disease which killed hundreds of foxes. Investigations at the Canine Research Station of the Animal Health Trust at Kennett, near Newmarket, have established that the disease is a virus allied to encephalitis hepatitis. Further progress is halted until the station can get hold of some healthy foxes.

The Masters of Foxhounds Association, which is collaborating with the station, has asked its members to supply fox cubs. Keepers all over the country are on the lookout for first litters.

Foxes suffering from disease have also been reported in Cheshire and Cornwall. One danger is that it might be transmitted in its virulent form to dogs.

Mystery of the 20,000 Bees that flew to death
John Winter 16.8.60

What killed 20,000 bees as they flew out in nectar-gathering formation a few days ago?

Captain Edwin J Tredwell, Hampshire bee-keeping instructor, says they were poisoned by a weed-killer sprayed on a field of kale, a quarter of a mile from their hive at the County Farm Institute, Sparsholt, Winchester. The bees flew over it to reach a patch of willow herb.

But Mr Stanley Magee, representing a leading chemical firm who made the spray, denied it was poisonous. At an "inquest" into the bees' deaths, called by Captain Tredwell yesterday, the two sides maintained their views.

Weed-killer samples were sent to Rothamstead Research Station, where tests had proved negative. But Captain Tredwell remains unconvinced.

The Red National hopes may be barred
Daily Mail Reporter 24.12.60

The three Russian horses entered for the Grand National in March may not run after all. Their entry into Britain is banned under a Government Disease Order made in September. The order follows an outbreak of African horse sickness, a deadly disease similar to influenza, which has ravaged parts of the Middle East. Under the Order horses from a long list of countries, including Russia, are forbidden to enter Britain.

No application for import permits has yet been made by the Russians. It is possible the Order may be relaxed before they need to apply. It was imposed after an investigation by United Nations veterinaries. A meeting of the UN veterinary branch will take place in Paris on January 15.

S-t-a-m-p-e-d-e plan for Britain
John Winter 9.1.61

Wild-west rodeo shows, complete with bucking bronco competitions and stampeding steers, may be seen in British agricultural show rings next year.

Officials of some leading regional and county shows are examining a scheme to bring over the Calgary Stampede (seen by the Queen and Prince Philip in 1959) for a summer tour. If costs and transport problems can be overcome, they believe the spectacle would be a winner in their campaign to win back lost customers. It would also solve, for one season at least, the shows' dilemma of finding a new top-of-the-bill spectacle which would blend with their agricultural character.

First approach was made by the Royal Norfolk, one of the most enterprising societies, whose show in July is Britain's biggest two-day event. Mr Harold Jeffery, Royal Norfolk secretary, said yesterday that he has discussed the project with the promoters of the Calgary Stampede.

"There are difficulties, and the cost would be prohibitive unless we can get sufficient shows to co-operate in planning a full-length summer tour," he said. "We would have to reinforce our ring fences into a kind of stockade to protect spectators. It would be 1962, at the earliest, before a tour could be organised."

Mr Jeffery added that the dwindling number of mounted spectacles made ring programmes more difficult to organise every year. Most shows stage musical rides and drives, by mounted regiments and police. This summer the Royal Norfolk will have the King's Troop,

Royal Horse Artillery. Total cost for two days – about £1,500.

BBC runs into farm trouble

John Winter 11.3.61

> 8.30 Saturday-night Theatre
> *A Home from Home*
> A play for radio by Lionel Brown

A radio play about life down on the farm, to be broadcast by the BBC tonight, has upset the farmers' union. The impression *A Home from Home* is expected to give about farmworkers' conditions is completely false, says the NFU. As a curtain-raiser to the play, its author, Lionel Brown, describes in the *Radio Times* three-roomed tied cottages lit by oil lamps and candles with hand-worked water pumps in the garden.

Last night the union issued a denial of all Mr Brown's claims. It was based on Government statistics and a survey made last year for Bristol University.

Said Mr Brown: Farmworkers have to start work at 6 am.

The Union: Average starting time is 7.0 to 7.30.

Mr Brown: Last year 35,000 of the "best young men and girls" left the country for cities.

Union: Between censuses in June 1959 and June 1960 the farm labour force fell by 24,000. One-third were women, seasonal or part-time workers.

The union also says that 93 per cent of tied cottages have piped water and 94 per cent electricity.

Mr Brown, who is 72, said at his home in Montford Bridge, Shropshire, last night: "I wrote the play after discovering the conditions in a group of 14 cottages not far from here." It was these cottages he described in the *Radio Times*.

He added: "My description is accurate, but I must say I wrote the play 12 months ago and conditions have improved rapidly since then."

TV weather switch 'too late for us'

John Winter 17.11.61

Farmers are in revolt against the BBC for switching the TV weather charts to 11 pm, when most of them are in bed. Not until they moved the time from 6 pm last month did the BBC discover that thousands of farmers used the charts to plan the next day's work.

Ten county branches and scores of individual members have protested to the National Farmers' Union. And last night a letter was sent from union headquarters to Broadcasting House. "Please can we have the detailed weather charts some time between 6 pm and 9.30 pm?" it asked.

"Many farmers have depended on the information they give, and when it was screened in the early evening they could organise their timetable for next day," said Mr Charles Jarvis, NFU publicity chairman. "At 11 pm it is far too late. Most farmers start work at 6 am, and we have suggested 9.30 pm as the latest time for the charts, because that is when farmers start going to bed."

The BBC has referred the pleas to its programme planners.

1962

A serious outbreak of fowl pest dominated the news for much of the year and, like the foot-and-mouth outbreaks in 1967 and 2001, was to continue for many months; swine fever and anthrax also made the news. The Royal Agricultural Society of England announced its intended move to a permanent site at Stoneleigh in Warwickshire that was to become The National Agricultural Centre. Grade 2/3 agricultural land was selling for £191 per acre and the workers' wage had still not reached £10 per week. The Common Market was the principal political topic and the arguments for and against marketing boards occupied the minds of many engaged in the industry.

 POLITICS and FOOD

One farmer and a boy challenge the Six   *John Winter 11.1.62*

Farmer Mr Geoffrey Spear startled Britain's agricultural "Brains Trust" yesterday by explaining how one man and his boy make "a decent living" from a 160-acre farm. It was due to following a plan of simplified farming suggested by Government farm business analysis experts, he told a farming conference at Oxford. If other smaller farmers would take similar advice, he said, they could look forward to competing successfully with Continental farmers if Britain enters the Common Market.

Mr Spear, 41, farms at Great Missenden, Buckinghamshire. He got the biggest "hand" of the two-day conference when he said: "I am a small farmer facing the future with confidence, making a decent living, and lining up my farm to meet the impact of the Common Market."

Before his father retired two years ago, the farm had a herd of 20 dairy cows, pigs and poultry, and employed four men. Now all the livestock have gone. The farm is devoted to growing herbage seeds and barley, with 36 beef cattle on a piece of unploughable permanent pasture. The staff: Mr Spear and a 15-year-old apprentice, helped by casual labour at harvest time.

Farmers say no to the Six: Common Market levy 'would raise prices and hit Commonwealth'   *John Winter 16.1.62*

British farmers believe that Sunday's agreement by the Six on farm policy presents no sound basis for Britain entering the Common Market. Their leaders say the conditions would be totally unacceptable to both British farmers and Commonwealth food suppliers. This is despite the shelving of price harmonisa-tion and the plan for a common agricul-tural fund. Part of this fund would be raised by import levies and would have two main effects:

1. Inevitably, if there is no modification, shop prices of Commonwealth food-stuffs would go up; and
2. Commonwealth exporters of food to Britain would be hit.

The levy means that a country outside the Market sending in goods at a price lower than that of home-produced prod-ucts would pay to the importing country a levy equal to the difference. Australia, New Zealand and Canada would all be hit. When negotiations on Britain's application reach food and agriculture this month, Mr Heath will probably insist on our right to maintain the quan-tity of our Commonwealth imports.

The National Farmers' Union said of the agreement that it was clear the Six had insisted on changes where the origi-nal proposals were likely to affect adversely their own farmers' imports. "The fact that these decisions represent a first stage in the evolution of a common agricultural policy to suit the Six them-selves in no way affects the union's general attitude to what it regards as the necessary requirements of a common policy for agriculture in an enlarged community, which would meet the vital interests of British agriculture and horti-culture," it added.

Tomato Board now aims to make a fresh start   *John Winter 27.1.62*

The much-criticised Tomato and Cucum-ber Marketing Board said yesterday that its 20 elected members – who were understood to have resigned on Thurs-day night – had resolved only to offer their resignations to their sponsoring bodies, the National Farmers' Union branches.

The union, which fathered the board, thought that they had resigned, leaving an interim executive to hold the fort, but both board members and the NFU are determined that the board shall not die. They want it to make a new attempt to win the confidence of tomato growers, who last month gave it a no-confidence vote and cut members' pay to 1*d* a year.

Said Mr Ken Ripley, 40-year-old tomato grower from Newton Abbot, Devon, who is chairman of the seven-man executive: "I believe the board can give valuable service to producers, and that its responsibilities will increase if Britain enters the Common Market. The board has decided to make a fresh start. Each member will offer his resignation to his sponsors who will either ask him to carry on, or accept it and nominate somebody else for election by growers."

The NFU is likely to urge acceptance of all resignations, which would cause a tomato board general election, even if some branches wish to re-nominate resigning members.

Fight over price cuts will be bitter   *John Winter 7.2.62*

The annual Price Review talks which open today between the Ministry of Agri-culture and the three farmers' unions are expected to produce the bitterest haggle since the system started after the 1947 Act. The Government is expected to demand maximum cuts.

It is faced with an excess estimate of £78 million, bringing the cost of farm support to £350 million against £265 million last year. The total permissible is

2½ per cent on the value of the guarantees, about £32 million, less increased costs – the farm workers' wage rise, rents and other items. This will bring the limit to less than £20 million.

The Ministry will have to meet the Exchequer axe, so it may seek other ways to contain the cost below £300 million next year. It may propose diverting some guarantee funds to strengthen the farmers' competitive position through marketing and co-operative schemes.

The guarantee for pigs is pledged for next year and there is also a conditional promise on beef. In spite of the extra £35 million it has cost this year, the unions will strongly resist any cut. This leaves cereals, potatoes, sugar beet, sheep and wool, milk and eggs, to bear the Ministry chopper.

Ministry cuts its losses on land

John Winter 9.3.62

More than half the land owned by the Ministry of Agriculture in assorted holdings throughout the country is up for sale. Mr Christopher Soames, the Minister, told Parliament yesterday that he is selling 74,000 acres of his 154,000-acre "estate". He is getting rid of a further 49,000 acres as soon as possible. This will leave him with only 31,000 acres.

The Minister's move follows losses on management and a recommendation by the estimates committee that he should sell out as quickly as possible. Last year the loss on farmland managed for him by the Agricultural Land Commission – 130,890 acres in 32 English and 12 Welsh counties – was £19,685.

Among early sales will be the 36,600 Glanilyn estate in Merionethshire, taken over by the Government in place of death duties. Talks for a collective sale are now going on with representatives of 102 farmer tenants. On the 49,000-acre list for later disposal are extensive mountain areas and other land suitable for tree planting which will be first offered for transfer to the Forestry Commission.

Land to be retained includes the Ministry's own experimental farms (11,488 acres); the Land Settlement Association's smallholdings estates (8,289 acres); the National Stud in Dorset (1,017 acres); and Kew Gardens (300 acres).

Common Market food and drink warning

John Winter 30.5.62

The head of a grain firm warned yesterday that food and drink will go up if Britain joins the Common Market. Mr Henry Nicholls, managing director of a Dunmow, Essex, firm of agricultural merchants, said that prices of whisky and beer, breakfast cereals and flour will rise. Three-quarters of all the country's other foods will be indirectly affected.

Mr Nicholls, retiring president of the National Association of Corn and Agricultural Merchants, told the annual meeting at Scarborough, Yorkshire, that the Common Market proposals did not match up to the present system. "The present arrangements which allow a free market in grain have worked well."

Strawberry prices up to 10s a pound

John Winter 15.6.62

Strawberries were selling at 10s a lb in the Vale of Evesham yesterday – the dearest they have ever been at this time of year.

The reason, said the growers in England's richest market gardening area, is that not a single outdoor strawberry is yet ripe. The crop has been held back by the long cold spring frosts and lack of sunshine, and the earliest beds are not expected to be ready for picking for another ten days. The fruit stacked on roadside stalls has all been grown under cloches. Although supplies are quite plentiful, the brisk demand, especially from holidaymakers, is keeping up the price.

Last year, with an early season, the outdoor strawberry crop was half-finished at this date, and the price was down to 2s 6d a lb.

Tomato Board woos rebel

John Winter 8.8.62

Mr Harry Wright has fought for 12 years to kill the Tomato and Cucumber Marketing Board. Today he is likely to become chairman of its publicity committee. The 47-year-old Yorkshire tomato grower, elected to the board on an abolitionist ticket, is still determined to wind it up.

On June 12 he was appointed to the publicity committee, which holds its first meeting today. The election of a chairman is the first item on the agenda. Six weeks ago the board said it took a serious view of statements by Mr Wright, and said he had been describing himself as chairman of the publicity committee. Mr Wright denied this and accused the board of underhand conduct in issuing a statement without consulting him.

The decision to nominate him follows off-the-record discussions between board members and leaders of the NFU horticulture section. Said Mr Wright: "When I first heard of it I thought it was a joke, because everybody knows I am pledged to work for the board's abolition. I have decided to accept the post, if it is offered to me, so long as it is clearly understood I stand by my principles to revoke the marketing board scheme as soon as possible."

Mr Wright, though determined to kill the board, with its compulsory levies, is ready to support a voluntary marketing scheme, run on co-operative lines. His support would be invaluable in putting this over to growers if the board were wound up.

Farmers oppose joining Market

John Winter 24.8.62

Farmers will oppose Britain's entry into the Common Market on the terms so far negotiated. Instead they urge the Government to seek a system of food agreements, embracing the Commonwealth, Europe and other temperate countries. Mr Harold Woolley, president of the National Farmers' Union, was speaking for more than 250,000 farmers when he launched their plan yesterday.

The new get-tough line has emerged from talks between leaders of the NFU (England and Wales) and the Scottish and Ulster farmers' unions. The 11-point alternative plan awaits formal approval by the 145 members of the NFU council next month, but copies will be sent now to the Government, MPs and Commonwealth Governments. It is based on a system of international agreements for the main food products.

Mr Woolley said: "We believe the Government should adopt a broad approach to the agricultural problem, rather than one based essentially on a part of Europe."

Main points of the plan are:

• The Organisation for Economic Co-operation and Development, already comprising the Common Market and the European Free Trade Association groups, with US and Canada as full members, and Japan as associate, should be enlarged to include Australia, New Zealand and Argentina, and create trading relations with the Soviet bloc.

• Annual reviews would provide commodity agreements.

• Horticulture, bedevilled by surpluses, would require a tariff system.

• International agreements would be consistent with Britain's GATT obligations.

• Britain should immediately join in a world food programme to build up buying power in undeveloped countries.

Heath wins the farmers over

John Winter 15.11.62

Not one dissentient voice was raised from a conference of nearly 1,500 farmers yesterday, as Mr Edward Heath gave them a one-hour progress report on Common Market negotiations. The farmers, at the Friends' Meeting House in London, gave the Lord Privy Seal a big ovation after his speech. Then he called for questions, and got plenty. But not one was hostile, and some were frankly enthusiastic about a European link-up.

He assured the farmers that he refused to negotiate against a deadline, although a wide area of uncertainty would exist, in Britain, the Commonwealth, Europe, the United States and many other countries, as long as the negotiations continued.

Mr Asher Winegarten, the National Farmers' Union chief economist, told the meeting that if Britain joined the Six "the rise in market prices would apply to all our food." He added: "Butter, of which we import about 90 per cent of our needs, could easily go up to around 6s a lb, and the price of cheese, of which about half our supplies are imported, could also go up by about one half."

CROPS and MACHINERY

Make more use of the sugar beet factories

John Winter 16.1.62

Idle sugar beet factories should be used for drying surplus potatoes to make alcohol motor spirit, says farmer John Dowty, of Ombersley, near Worcester.

Spring freeze-up hits prices: wind and frost put a blight on early crops

John Winter 26.3.62

Vegetables, salads and flowers will soon be scarce – and dear. Bitter winds and severe night frosts, persisting for over a month, have played havoc with farmers' early crops. Seed potatoes have been frozen and killed in the ground. Spring greens stand scorched and shrivelled. Salad crops are at a standstill. Fruit trees are weeks behind in blossoming. Spring flowers are late and stunted, and grass for early grazing is almost non-existent.

From all parts of the country yesterday came reports of crops, which were sown and planted in record time, now lying dormant in the cold dry earth.

Potatoes are the most serious casualty from the housewife's point of view. Last year's crop, below normal in weight, and deteriorating in quality, is expected to run out by early May. The new Jersey crop, which usually follows, has been set back by hard weather in the island, and the home crop will also be late. A gap is likely which can be filled only by expensive imports of new potatoes from the Mediterranean, or old ones from Europe.

Spring greens have been badly frosted. Supplies will be short, and poor in quality.

Salads: Lettuce in heated glasshouses will be a money spinner, but cold house crops are late, and outdoor ones have suffered serious damage.

Flowers: Outdoor daffodils have been badly retarded, and prices for Mothering Sunday bunches next weekend are likely to be the highest ever.

Fruit: Orchards which were in full bloom in mid-March last year show no sign of blossoming. Strawberry plants in the Cheddar Valley have been blackened by frost.

Grass is the most serious shortage for farmers. There is no sign of the "early bite" they try to produce for their cattle and sheep.

Rain brings hope of cutting the potato famine   *John Winter 7.5.62*

The Potato Marketing Board has had its best news since it admitted on April 11 that supplies were running out – weekend rain in the farmlands which grow early crops. If it is followed by sunshine, the gap between the end of high-priced old potatoes in the shops and the arrival of new ones from our own farms will be shortened.

But this is the only ray of light. A board team returned from Jersey at the weekend after checking on the progress of the island's potatoes with the grim report that they will not be ready in quantity until next month. Last year they were ready from mid-May. Unless the earliest home crops in Cornwall and

Pembrokeshire have ideal conditions they will be ready at least a fortnight later. On top of this the board is still not sure that enough potatoes have been planted to avoid another shortage on this year's crop.

After securing sanction to raise the penalty for exceeding quotas from £10 an acre to £25, it has decided to charge no excess levy at all. But it is doubtful whether this will encourage farmers to plant more. Board vice-chairman Mr H P Renwick said: "The main problem is the high price of seed potatoes. Seed which usually costs £30 a ton is £55–£60 today."

Forecasts based on the March returns showed only a small increase of potatoes in England and Wales and the early crops figure was actually 10 per cent down on last year. The total farmers expected to plant was 527,000 acres. At the same date last year, they returned 508,000 acres. In 1960, which produced a surplus, it was 598,000 acres.

ICI cuts fertiliser costs for farmers   *John Winter 21.5.62*

Eleven weeks after winning a £3-a-ton anti-dumping duty on sulphate of ammonia from East Germany, ICI today announces an 8s 6d-a-ton reduction for its own product. All other fertilisers it sells to farmers, in many of which sulphate of ammonia is a major ingredient, are cut by 10s to 15s a ton.

The firm claims this will save farmers nearly £1 million on their fertiliser bills for next season. Similar cuts were made in each of the last two years.

Until the anti-dumping duty was put on, the East German sulphate was underselling ICI in Britain. Mr Rudi Sternberg, leading advocate of trade with

Eastern Europe, and a big farmer in Kent who imported 30,000 tons a year, alleged that high sulphate prices in Britain were due to inefficient manufacture.

ICI says the price reductions, made in spite of increased cost in wages and services, have been made possible by more efficient production and distribution, and by anticipating savings from new ammonia plants.

Storms spell harvest disaster

John Winter 8.8.62

A heartbreak harvest scene of tangled flattened corn crops greeted farmers yesterday after torrential rain and gales. The gales cut a swathe through the richest corn country, from Lincolnshire to Kent, and delayed still further a harvest already three weeks late. Western counties escaped lightly. Many eastern wheat and barley crops were totally flattened, meaning losses worth thousands of pounds, for the crops are not yet fit for cutting and will not ripen properly.

Essex farmers are talking of a repetition of the disastrous 1958 harvest, when corn was gathered up to Christmas and some crops were never harvested. In Kent orchards also suffered and in West Sussex one-third of the corn is down.

'Eye in the sky' forecasts a month of rain   *John Winter 17.8.62*

Another month of rain and cold. That is the forecast of American experts as the worst August for 30 years begins to threaten Britain's corn harvest.

The prediction comes from the US Weather Bureau, using experimental Tyros satellites to keep a long-range track on weather round the world. From now until mid-September it forecasts below-normal temperatures for all the British Isles, much below normal in Scotland and Northern Ireland – and heavy rain. As holidaymakers sheltered from the rain yesterday British forecasters admitted that the American weathermen could be right. The holiday industry is having one of its worst years ever, with a slump in bookings and the sales of ice-cream and soft drinks.

If the forecast proves correct it would also mean a shattering blow for farmers. Already harvesting is three weeks late. In the southern corn belt, from Sussex to Dorset and north to Berkshire and Oxfordshire, a fifth of the nation's grain stands ready for cutting in the next three weeks. The crop is good. A dry spell would give many farmers as good a harvest as they have ever reaped.

Crops of all cereals recovered well from the long cold spring and have suffered remarkably little damage from gales and heavy rain this month. The amount of flattened corn is negligible. "If the harvest can start now and continue without interruption I believe we shall get even more grain than last year," said Mr John Loader, executive of a Hampshire firm of corn merchants.

Big farmers with upwards of 500 acres of cereals are not so far unduly perturbed by the hold-up. With two or more combines apiece, and grain dryers and stores on their farms, they believe they can still get a good harvest provided it is not too long drawn out.

For holiday resorts, their profits already halved by the weather, another month's rain would be the final blow. It would make this a year without a summer. Much of Southern England and the Midlands has had less than three-quarters of the normal sunshine this month. Parts of Devon have had less than half. All along the South Coast and inland rainfall has been double the average.

Apple acreage *John Winter 21.8.62*

Sale of 600 acres of apples and pears at Horsmonden, Kent, realised £80,000.

Shock news for wide-row sowers *John Winter 4.9.62*

Growers who cling to the old idea that vegetables need plenty of room to grow had a shock yesterday. A physiologist showed that the theory was wrong.

Dr J K A Bleasdale, head plant physiologist at the National Vegetable Research Station at Wellesbourne, Warwickshire, was addressing the horticulture section of the British Association at Manchester. He said that higher yields of carrots and onions had been produced from seeds in rows sown close together than from crops sown in widely spaced rows. Rows of carrots four inches apart had given 47 tons of marketable roots per acre against 34 tons from plants grown in rows 16 inches apart.

Dr Bleasdale said: "If future research continues to show with other vegetables that close rows can give higher yields, engineers will have to design suitable sowing and harvesting machinery before close rows can become a commercial success. Then it is envisaged that the method of growing will resemble the bed-system of cultivation used in market gardens of the eighteenth and nineteenth century, when hand labour was cheap, except that the tedious and expensive hand labour of today will be eliminated."

An end to the drifting poison

John Winter 31.10.62

The deadly drift of chemical spray, which hangs like a poisonous cloud over farmlands, has been conquered by a simple new machine which was tested in Sussex yesterday.

For only £135 farmers can buy an 18ft sprayer for mounting on their tractors, which will put weed-killing liquids exactly where they want them without any danger of drift on to hedgerows, neighbouring crops or gardens. The principle, which will save thousands of claims for damages every year, is to shake the liquid through holes in droplets too large to be carried on the wind instead of forcing it in a mist through nozzles.

The sprayer has been devised primarily to apply the revolutionary new chemical, Paraquat, to reconstitute grassland without ploughing, and other selective weed-killers. Prototypes were built at the firm's 500-acre research estate at Fernhurst, Sussex, and passed to farm machinery makers four months ago. First production models will be displayed to farmers at the Royal Smithfield Show in December.

 # LIVESTOCK and POULTRY

Quarantine bulls freed

John Winter 21.1.62

Twenty-four Charollais bulls, in quarantine at London Docks for 11 weeks after three others were found to be diseased and slaughtered, were released yesterday and sent throughout the country for breeding tests.

Two more have become lame and will be treated before going out. A third leaves for the Isle of Man today.

Egg team gets cracking: they will smash 4,000 in test

John Winter 29.1.62

A team of experts will start work today smashing 4,000 eggs. As each egg is broken they will test its colour, protein value and shell thickness. Object of the big smash at the *Daily Mail* laying test ground at Milford, Surrey, is to find out which eggs are good eggs. The researchers, who are from the Egg Marketing Board, hope that their records of each egg will answer such questions as:

• Does egg quality vary from one breed of chicken to another as milk does in different cattle breeds?
• Does the big drive by breeders to produce cheaper eggs affect their quality?

In all, eggs from 43 different breeds sent by breeders all over Britain will be tested. Each egg will be weighed and examined for faults, such as bloodspots on the yolk. A micrometer gauge will measure how much there is of the thick, jelly-like substance near the yolk, for this has the highest food value.

Dr N R Knowles, the Egg Marketing Board's chief scientist, said: "We are constantly searching for more information about factors which affect egg quality. The laying test gives us an excellent opportunity to get a broad cross-section of eggs laid by hens of different strains and crosses which have come from all parts of the country. We are watching especially for any effect on quality caused by the constant drive by breeders to get higher and more economical egg production."

The big smash will last all week; but not an egg will be wasted. After recording they will all be processed and frozen, for use in bakeries.

Free of disease

John Winter 8.3.62

Livestock in Devon is free from disease except for two cases of anthrax in cattle, says divisional veterinary officer, Mr E Addison.

What puts the gambol in Mr Williams's lambs . . .

John Winter 23.3.62

Those lambs you'll see skipping behind a hundred hedges when you drive through the country this first weekend of spring are not so daft as you may think. Nor are their mothers, grazing watchfully nearby, the silliest creatures on the farm, as everybody has said since the Bible was written.

It has taken farmers 2,000 years to learn that a sheep is a wise animal, responding to certain predictable impulses. Now they are beginning to realise that a study of sheep psychology does not only explode a lot of misconceptions, it pays off handsomely. I have just investigated the results of ten years' study of sheep psychology by one of Britain's best-known farmers – Mr Stephen Williams, general manager of 20,000 acres of farms in England and Scotland – for Boots the chemists. And what has Mr Williams discovered?

1. That if you put more than 80 ewes – 72 for absolute safety – and their lambs in a confined space you're heading for trouble. Much like mother and daughter-in-law using the same kitchen in fact. Except that with sheep

they develop suicidal tendencies and mass hysteria.

2. That if you want to make money as a sheep farmer you've got to *teach* the newly born lambs to eat grass as fast as they take their milk . . . until they're chewing steadily from dawn to dusk.

Mr Williams's experiments have all been made on the 1,000 acre Thurgarton Estate in Nottinghamshire. Having found out how many sheep could live happily and profitably – for him – together, he next set about finding how much room they needed and whether or not they liked neighbours. Answer: eighty ewes and their lambs need a fresh nine acres to graze every 16 days.

However, the grazing has to be scientifically arranged. The lambs must crop the grass before the ewes set foot on it. So the nine-acre plot is divided into six 1½-acre paddocks. Between each paddock fences are arranged so that the lambs can feed on the grass adjacent to the patch their mothers are cropping.

But before the lambs are turned into the paddocks Mr Williams discovered that he must teach them to have sharp appetites so that they will eat, eat, eat! How did he do this? Simply by playing a confidence trick with mother-love.

Before the lambs are turned into the paddocks Mr Williams implants in them an indelible idea that mother always – but always – knows best. This takes eight days in a nursery, where the lambs are continuously with their mothers. It fixes her function firmly in their minds, so they will never again have any difficulty in knowing her and will always copy her. Which plays right into Mr Williams's hands to get them eating vigorously.

Once in the paddock he ensures that the ewes are always reasonably hungry, and have to eat hard at the grass which dwindles rapidly as the day approaches for the move to the next paddock.

Mr Williams also discovered that when it comes to neighbours sheep are rather like the southern English. They don't like them. So the whole 9-acre field is enclosed in a closely grown "A" shaped hedge, which deflects outside noise upward, and all gates are boarded.

And don't worry when you see those lambs gambolling instead of grazing this weekend. They do it because they're bouncing with health, not hysteria. Don't worry either if you see lambs lying down. Even on the farm there can be a black sheep in every family.

Outbreaks of fowl pest

John Winter 28.3.62

9,500 head of poultry have been destroyed in Norfolk because of two more outbreaks of fowl pest. Mr A Bailey and Sons, Hill Farm, Ashby St Mary, lost 4,500 and Mr F Burrows, of Scoulton poultry farm, lost 5,000.

John Winter 29.3.62

7,380 head of poultry were destroyed in Norfolk yesterday following six further outbreaks of fowl pest.

New fowl pest scheme brings in protests

John Winter 30.3.62

The system of fighting fowl pest, Britain's most expensive farm disease, by total slaughter and compensation will be abandoned if a Government committee plan is adopted. Instead, only birds actually suffering from the disease, and other prime suspects, would be slaughtered. Other contacts could be saved by prior vaccination.

This would save the taxpayer the cost of compensation, which has totalled £17 million in the last seven years, but the poultry farmer would be saddled with the bill to cover the cost of the new system, except for veterinary services, slaughter and disposal of birds. He would have to pay for his own vaccination costs, which might be as high as £3 for 1,000 shots, and birds would be jabbed as chicks, at six months, and thereafter annually.

Compensation for birds it was decided to slaughter, even though they showed no symptoms, would be paid out of a levy on chicks, turkey poults and ducklings produced by hatcheries. (No estimate of cost can be made from past figures, since 99 per cent of the 16,500,000 birds destroyed since 1954 were apparently healthy, but sentenced under the policy of total slaughter of all stock on an affected farm.)

Protests at this new "blow to the industry" started to pour in from poultry organisations as soon as the committee's report was issued yesterday.

The Egg Marketing Board: "Coming on top of the price cut for eggs in the annual review the recommendations appear quite unjustifiable."

The British Turkey Federation: "The recommendations present a very serious additional expense to our industry."

The National Farmers' Union: "We are forced to the conclusion that the only effective way of implementing the slaughter policy – the most effective measure of control – was through the provision of Government compensation as at present."

And there were more protests at the committee's proposal that its plan should start in the new financial year, beginning next week. Said the Turkey Federation: "It is preposterous."

The basic conclusion of the committee, set up in 1960 under the leadership of economist Sir Arnold Plant, is that it is no longer feasible to fight fowl pest by an eradication policy. This implies acceptance that the disease is endemic in Britain – as in many other countries including the US. It would also leave foot-and-mouth disease as the only scourge which is at present tackled by eradication methods.

Last night the Ministry of Agriculture said it would discuss the Plant scheme as soon as possible with all sections of the industry.

John Winter 5.4.62

Fowl pest outbreaks at Ickburgh and Sparham, Norfolk, have resulted in 5,300 poultry being destroyed.

Two-year plan to end fowl pest

John Winter 24.4.62

A plan to wipe out fowl pest has been suggested by the Norfolk branch of the National Farmers' Union. The branch executive says compulsory vaccination should be carried out in an area of the country over two years at the Government's expense. The present slaughter and compensation policy would stay – and poultry imports would be banned.

Yesterday 79,000 head of poultry were destroyed in Norfolk following another 18 outbreaks of fowl pest. Nearly a million birds have been killed in the past seven months.

100,000 turkeys killed as fowl pest strikes By Daily Mail *Reporter* 21.5.62

White-coated Ministry of Agriculture workmen began slaughtering 100,000 turkeys on a farm at the weekend because of fowl pest.

The birds were reared by Mr Bernard Matthews of Great Witchingham, Norfolk, Europe's largest turkey producer. For the third time in 18 months fowl pest has struck his farm at Weston Longville, bringing the total loss after the weekend's count to more than 400,000 birds. This is the greatest single fowl pest disaster ever to hit the turkey industry.

Mr Matthews, who has another farm at Langham and 40 associate farms, said: "It is time the Government abandoned its senseless slaughter policy. I think vaccination is the only cure – something we are not allowed to do. The present policy is like trying to eradicate the common cold by shooting every man, woman and child who catches it."

John Winter writes: *At present all birds on an affected farm are slaughtered, although only a small percentage usually shows symptoms of the disease. Compensation is paid on all apparently healthy birds.*

A few shell shocks

John Winter 28.5.62

Big Egg Smash No 2 will start next Monday when a team of egg cracking experts get to work at the *Daily Mail* laying test ground at Milford, Surrey.

The experts will smash and examine more than 3,000 eggs from the test's 64 pens of commercial hens. But what puzzles the men who bred the hens is that the boffins from the Egg Marketing Board have not yet told them what they learned from Big Egg Smash No 1 which they carried out in January.

Aim of both smashes is to find which eggs are the best, judged by weight, yolk colour, protein value and shell thickness. Breeders have been pestering both the board and the laying test office to find how their birds fared, but so far they have not learned a thing. Said Dr N R Knowles, the board's chief scientist: "One reason is that the final test on the January operation is only just finished. This was the measurement of shell thickness." But there are other reasons, which may lead to shocks for some breeders.

The eggs tested next week will be laid by the same hens as those smashed in January, and each egg is meticulously recorded back to its pen. At the first smash the hens, which started to lay last September, had just hit peak production. Now they are tailing off. Eggs, however, are bigger. In January only 34 per cent of all those laid in the 64 pens were in the large grade. Now the level is 60 per cent.

But some hens decline more quickly than others, not only in the number of eggs laid, but in their internal quality and shell strength. The second test will sort out the consistent hens which stay the course and continue to lay good eggs when prices rise in the summer from those which start flagging.

Dr Knowles said: "Our findings will be published, and detailed information on their own birds will be available to breeders, but we thought it would be more useful to say nothing until we can give them a complete picture."

The smashed eggs will be processed and frozen for use in bakeries.

Swine fever outbreak
John Winter 1.6.62

Pigs – nearly 700 – have been destroyed on farms near Bath as a result of swine fever.

Mechanical shepherd
John Winter 12.7.62

Mechanical shepherd invented by Mr B W Aston, of Chilham, Kent, comprises a large wooden box on a buckrake to lift newly lambed ewes.

Milking and farrowing productivity
John Winter 13.7.62

New milking parlour at South Lopham, Norfolk, will allow nearly 200 cows to be milked under the charge of two men instead of ten.

John Winter 13.7.62

Sow on Mr John Pegram's smallholding at Clipston near Market Harborough, Leicestershire, produced 19 piglets; another had 17.

Now it's ½d jabs for fowl pest
John Winter 19.7.62

A new fowl-pest policy to supply poultry farmers in England and Wales with ½d-a-dose vaccine over the next two years was announced yesterday by Mr Christopher Soames, Minister of Agriculture.

It means that the present slaughter and compensation scheme which is costing the Exchequer £10 million a year will be scrapped. The reason: slaughter is not wiping out the disease. But compensation will continue throughout the whole country until March 31 next year, while the manufacture of the vaccine is getting under way and stocks are being built up. Under the new plan the Government will subsidise the cost of the vaccine for the first two years. Farmers will be able to give the vaccinations themselves from supplies obtained through Ministry divisional centres.

At a Press conference yesterday Mr Soames said he had accepted the advice of the Plant Committee, which had concluded that slaughter was not succeeding. The success of the new plan depended on a very high proportion of birds being vaccinated.

In the Commons earlier Mr Soames said in a written reply that since slaughter was unlikely to eradicate fowl pest, the Exchequer could not go on indefinitely bearing the heavy cost of compensation, simply as a means of keeping the disease under control. "I have, therefore, decided to subsidise the price of vaccine for two years and to discontinue at the end of the present financial year the present policy of slaughter, with compensation at Exchequer expense, except for the rare peracute form of the disease," he said.

Last night a National Farmers' Union spokesman said the union fully supported the vaccine scheme. He added: "The policy introduces, for the first time in England and Wales, a new element in fowl pest control – the use of dead vaccine."

Charollais bull killed by rare disease
John Winter 26.9.62

A rare disease called "black quarter" has caused the fifth death among 30 Charollais bulls, imported from France last November for beef breeding tests.

The bull, Signor, nearly two, died suddenly at the Milk Marketing Board's centre at Chippenham, Wiltshire, last week, but the cause of death was not suspected until a post-mortem report was completed yesterday. Of thousands of bulls which have passed through the board's AI service, he is the first casualty to this disease. His death is not causing alarm, nor have any special precautions been taken.

A Ministry of Agriculture spokesman said: "Black quarter, which can attack any livestock, is a germ which gives rise to a septic condition and kills very quickly, but it is usually confined to one animal and does not spread."

Lancashire outbreak

John Winter 27.10.62

Fowl pest in Lancashire is causing Ministry of Agriculture officials "grave concern". There have now been nearly 200 outbreaks. In the Ormskirk district alone, 30,000 birds have been destroyed in a fortnight.

Losses in Norfolk *John Winter 14.11.62*

Poultry farmers in Norfolk have lost 3,500,000 birds in 1,500 fowl pest outbreaks in the past two years.

At a meeting organised by the Norfolk Farmers' Union in conjunction with the Norfolk fowl pest committee, farmers were told that since the Ministry made available its subsidised vaccine for protection, 244 applications had been received and 750,000 doses distributed.

Veterinary officers estimate that they have nearly enough vaccine to protect 2,500,000 birds – about half the poultry in Norfolk.

From midnight tonight all parts of Norfolk and Suffolk will be free from fowl pest restrictions for the first time since March.

Burning Question

John Winter 27.12.62

Scientists have started work on finding the answer to the burning question of how well the French Charollais bulls compare with those of British beef breeds when used for crossing with dairy cows. The first calves sired by the 24 Charollais bulls imported by the Ministry of Agriculture and used at the Milk Marketing Board's artificial insemination centres are now being born on farms all over the country.

Detailed records are being kept of about 1,200 Charollais crossed calves on Ministry, university and Farm Institute holdings. On commercial farms a maximum of 3,400 Charollais crossed calves and 3,000 crossed calves by British beef breeds will be compared, using ordinary farm records.

This adds up to the biggest, most comprehensive and most expensive cattle-breeding experiment made in Britain.

EMPLOYMENT

'Left behind' farm men put in 25s claim

John Winter 2.7.62

A 25s-a-week wage claim by 500,000 farm workers will be lodged today.

The National Union of Agricultural Workers, which will place the applica-

tion before the Agricultural Wages Board, will stress that men and women on the land are falling further behind other workers. Although agricultural workers had a 6s rise in February, which brought the minimum wage for a man to

£8 15s a week, they have since lost ground in the wages "league table". The claim is the first for several years in which the workers have asked for a specific amount.

The NFU, when called on to reply, will argue that the industry cannot afford the increase.

An extra 8s a week down on the farm
John Winter 21.9.62

Farm workers are to get an 8s a week rise, bringing their basic wage to £9 3s. They had asked for a £10 minimum. The new rates will come into force on November 26. A 6s rise was also paid last February.

The decision – made unanimously last night after a 4¾-hour meeting of the Agricultural Wages Board – now has to be formally ratified on November 7. There will be proportionate increases for younger male workers, women and part-time and casual workers. The new rate means a basic 4s an hour and overtime at 6s an hour.

Mr H Collison, general secretary of the National Union of Agricultural Workers, said after yesterday's meeting that suggestions made at first had not been satisfactory to the workers' side. But after long negotiations the employers had offered the 8s. The basic working week remains at 46 hours. The union did not apply for a reduction, but Mr Collison has said that application will be made eventually.

Reaction from the National Farmers' Union: "This will mean an increase in the cost of production of review commodities of £9,750,000 in a full year. While this will be taken into account at the annual price review, it is by no means certain that farmers will be reimbursed."

Death rate down, but still too high, says Ministry
John Winter 2.11.62

The safety record on farms has improved in the past year. Last night, only a day after publishing a bleak report recording that an average of 120 men, women and children were killed on English and Welsh farms during the previous four years, the Ministry of Agriculture revealed that the death rate is down.

In the year ended in September there were 119 accidental deaths, against a peak figure of 138 in the previous year. Most gratifying was the drop in the number killed by overturned tractors – 36, against 43 – in spite of more machines being in use.

The Ministry has recently stepped up its campaign to spread a safety code for tractor drivers. Its most telling piece of propaganda, a film showing how accidents happen, is now being shown to farming audiences all over the country. It shows that most tractor upsets are avoidable, and illustrates graphically some of the commonest causes – too tight turns, especially on steep gradients; unwieldy loads which can pull the tractor over, and loaded trailers running away.

"The number of accidents is still far too high," said a spokesman. "The Ministry is anxious that all tractor drivers should learn how to avoid these dreadful accidents."

Overtime pay for workers
John Winter 21.11.62

The week beginning Sunday, December 23, will include two public holidays, Christmas Day and Boxing Day, and for agricultural workers overtime will be payable for any work done on those days and for any work done over 29 hours during the remaining three and a half days.

Daily overtime provisions do not apply to weeks containing public holidays so that for this particular week overtime must be calculated on a weekly basis. These arrangements apply to all of England and Wales with the exception of Northumberland.

PEOPLE in the NEWS

Farmer Smith lays siege again

Daily Mail *Reporter 16.2.62*

Pig farmer Ernest Smith, who squatted for three days in November at the Ministry of Agriculture in London, resumed his siege yesterday.

He arrived with shaving kit and pyjamas, prepared to sit in the Ministry waiting-room until Mr Christopher Soames, the Minister, intervenes in his dispute with his landlords, the Land Settlement Association. His last squat won a promise that the association would give a new hearing to his claim of inefficiency and mismanagement of its boar stud scheme at Harrowby Hall Estate near Grantham, Lincolnshire.

Mr Smith, who runs a ten-acre smallholding there, says he lost £260 when 12 sows he sent to the stud in 1959–60 produced very few piglets. He said yesterday: "I consider the association has not fulfilled the promise made through the Minister, and I shall stay here until I get satisfaction."

By the afternoon the only Ministry staff who had spoken to Mr Smith were two security men. He was told Mr Soames was not available. "I shall stay in town overnight and return in the morning," said Mr Smith. "As the Ministry does not work on Saturday, I shall go home for the weekend and come back on Monday."

Third generation
John Winter *13.3.62*

At 18 Mr Geoffrey Toone is the third generation of his family conducting cattle auctions at Market Bosworth, Leicestershire.

The hobby of an affluent society
Daily Mail *Reporter 30.4.62*

Farmer Edgar Weaver went to police about it when the matter reached sizeable proportions. Would they please protect him against the gardeners of Stoke Gifford (pop. 1,264), Gloucester?

These people, he pointed out, were trundling wheelbarrows and buckets from field to field on his Perrinpit Farm, following his cattle and sheep about, and collecting the by-products. Result: The output of by-product available for the farm itself was seriously curtailed. That, said Mr Weaver, meant that the gardeners in effect, were robbing his land of its fertility. A serious matter.

The police, however, were doubtful. It was certainly a nice point of law; indeed, it was new to them, but would a prosecution be a success?

The farmers of Gloucestershire are made of sterner stuff. Their branch of the National Farmers' Union issued a stern warning yesterday that *they* at any rate were willing to take a chance. *They* would prosecute anyone caught following this hobby.

Mr Weaver traces his troubles to the day a council estate was built near his land – to the theory that there is nothing like that by-product to give plants a good start in a new garden. He said yesterday: "Some go to enormous trouble. Why, it takes about an hour of hopping about to fill a bucket. They think they have a right to it, as they do to mushrooms, because it's scattered about the fields."

Comment from Lieut.-Colonel D G B Duff, the NFU county secretary: "Dung stealing (*please excuse his farmer's forthrightness*) has become a hobby in the affluent society. Some of these people are in the business in a big way – one has collected several hundred-weight."

Farmworker candidate

John Winter 24.5.62

Farmworker Mr Robin Martlew, 30, of Stanningfield, near Bury St Edmunds, Suffolk, has been adopted by the Bury Division Liberal Association as prospective Parliamentary candidate.

Farmer fights market scheme

John Winter 11.6.62

Farmer Richard Wheelock has launched a one-man campaign to stop the Fatstock Marketing Corporation, founded by farmers to sell their meat, being turned into a public company. The plan, announced by the FMC ten days ago, includes the £7,450,000 takeover of Marsh & Baxter, Britain's biggest bacon and sausage firm. It has received the blessing of the farmers' unions.

But Mr Wheelock, 42, of Wonastow, near Monmouth, says it means taking control of a business handling nearly £100 million of meat a year out of the hands of farmers, and handing it to "Stock Exchange manipulators". He said yesterday: "A commercial enterprise which aims to buy in the cheapest market, and sell in the dearest, is quite incompatible with our interests as producers."

The deal requires approval from a FMC meeting on June 26. Mr Wheelock is appealing to the corporation's 94,000 farmer members to authorise him by proxy to turn it down. Under the proposed new constitution the NFU Development Trust would hold 35 per cent of the equity in FMC, while 65 per cent would be on public issue, with NFU and FMC members getting preference in allotment.

The NFU says the 35 per cent stake is sufficient to safeguard farmers' interests, but would like to see many of the remaining shares taken by them.

Reunion means home for the lad

John Winter 12.7.62

A touching reunion has brought an end to a triumphant association for Mr John Howlett, 79, Grand Old Man of the show ring, and his big swaggering bull, Malverley's Robert's Lad. The bull, 12 in October, has won 55 championships in five seasons. He was paraded for the last time yesterday at the Great Yorkshire Show, Harrogate. He came second.

Mr Howlett is giving the bull, free, to Mr Chellew, the Cornish farmer from whom he bought the Lad for 500 guineas in 1958. Mr Chellew recently visited Mr Howlett's farm at Lymington, Hampshire. "The bull showed at once that he recognised his old master, and I decided he must go back to his old home," said Mr Howlett.

New Man at the Royal

John Winter 18.10.62

A civil servant has been appointed to one of the top jobs in British farming. He is Mr Christopher Dadd, 45, regional crop adviser with the Ministry of Agriculture in Cambridge, who becomes technical director and secretary of the Royal Agricultural Society of England. Mr Paul Osborn has been appointed deputy secretary and administrative officer.

The appointments, says the society, are a further step in the policy changes it is taking in recognition of the challenge facing British agriculture.

Bulls in bowlers may strike on pay

John Winter 22.10.62

A strike by the men who operate the Milk Marketing Board's artificial insemination service is threatened. The inseminators – known to farmers as "bulls in bowler hats" – earn from £10 14s to £15 10s a week according to service and grade. They want a substantial increase. The Union of Shop, Distributive and Allied Workers, to which 8 per cent of the inseminators belong, put in a claim in May and later rejected the board's offer of a 5 per cent rise. After further talks the board offered to cut working days from 12 to 11 a fortnight.

The union will this week receive the inseminators' verdict on whether to accept or reject this offer, refer the dispute to arbitration or terminate their agreement with the Board.

The board's artificial insemination service is responsible for the births of 1,700,000 calves a year in England and Wales. If the men cease work, the country's milk supply, which depends on a massive programme of carefully timed breeding, would quickly be disrupted.

Mr James Hughes, USDAW national officer, said yesterday: "When AI started it was an experiment, and sights on pay and hours were set low. It has proved a tremendous success and we feel that a completely new wages structure and overhaul of working conditions are overdue."

The board claims that because of wide seasonal and regional variations in farmers' demands on the AI service, considerable elasticity in working hours is unavoidable. Inseminators at five independent AI centres, whose pay unusually follows that of board men, are awaiting the outcome of the dispute.

Turned down

John Winter 31.10.62

The "Bulls in Bowler Hats" – the men who operate the Milk Marketing Board's artificial insemination service for farm cattle throughout England and Wales, have rejected the board's offer of 5 per cent wage increase to settle their pay claim.

This is the second time they have turned it down in negotiations which started in May. Unless the board is ready to offer more, the dispute can go to arbitration, or the "Bulls" can end their agreement, which would leave them free to strike.

GENERAL

Jumping for big prizes

John Winter 20.1.62

Show jumping for big money prizes put up by commercial companies will be introduced to agricultural shows this year.

Leading the way is the Royal Show at Newcastle-on-Tyne, in July, where Vaux, the Sunderland brewers, are offering £1,000 prizes for one contest. Bath and West Society also announced big prize-jumping competitions.

Big drive against wood pigeons is coming
John Winter 2.2.62

Britain is to make a big drive to cut down the numbers of wood pigeons – birds which cause a lot of damage on the farms. Mr Fletcher Vane, Ministry of Agriculture Parliamentary Secretary, told Parliament yesterday: "During the coming weeks, concerted shoots can be very effective."

He urged farmers, landowners and shooting people generally to give all possible support to the widespread shoots being organised. "Apart from many district and country-wide shoots, the counties in East Anglia are combining today and tomorrow for shooting over the whole region," he said. "Neighbouring counties will join in. A special shooting day is to be organised in Yorkshire today, and in Shropshire a series of shoots has been arranged for Thursdays in February and early March."

Land issues
John Winter 14.2.62

Eight acres of agricultural land within Norwich Green Belt area are not to be used for housing, says the Ministry of Housing and Local Government, following a public inquiry.

John Winter 23.2.62

160 acres of bulb and agriculture land at Gedney Broadgate, Lincolnshire, have been sold for £34,600.

Townsfolk give farmers pat on the back
John Winter 21.2.62

What do townsfolk really think about farmers? The National Farmers' Union has commissioned a team of experts and this is what has been found out so far:

Most of the people interviewed in London, Birmingham, Ipswich and Reading think a farmer works long hours, seven days a week in bad weather and good, and is not over-rewarded. Farm subsidies have made little impact on townsfolk, but they are vitally concerned about the price of food. The rumpus over meat estimates did not register at all, even with those interviewed while it was raging.

"It is too early to sketch the complete image: only one-quarter of the interviews are complete," said an NFU spokesman. "But what we have learned so far is distinctly encouraging." Hundreds of two-hour interviews were held.

The day whistle pop rocked the hens
John Winter 3.3.62

The trad-fad hens of Golding Hop Farm were delighted to have music while they worked.

Until this week when, as a special

treat, their owner Mr Eric Weston, of Plaxtol, Kent, switched on a new tape called *African Whistles*. The effect was shattering. The 2,500 birds started flapping their wings and their squawks drowned the recording.

Poultry foreman William Birtles said: "I can't understand it. We started giving the hens music three months ago and they laid better. Continental orchestras and Latin-American airs went over big. But this whistle stuff really upset 'em. I must admit I didn't much care for it."

The recording company said: "It probably has something to do with the frequency at which hens receive sound."

Boys do the ironing

John Winter 7.3.62

Three boys, members of the Northamptonshire Young Farmers' clubs, are the first in the country to pass proficiency tests in laundry work. They took a six-week part-time course in washing, ironing, starching and stain removing.

Killer dogs slaughter 77 sheep in field

Daily Mail *Reporter 15.3.62*

A pack of killer dogs which slaughtered 77 sheep in one night was being hunted by police yesterday. The sheep belonged to 28-year-old Mr Idris Powell, of Wernheargest, Erwood, near Builth Wells, Brecon. He was not insured and faces a loss of more than £1,000. All the lambs born to ewes which survived the slaughter have been born dead.

He started farming three years ago and had a flock of 165 before the killers swooped on one of his fields. He was expecting 230 lambs to be born this spring.

The attack was the second in South Wales in a week. Near Llandrindod 30 ewes were killed and others mauled.

Farmers believe that the dogs team up in packs to hunt for food after being abandoned by their owners in the lonely hills.

Yesterday the Welsh Committee of the National Farmers' Union, meeting in London, demanded stern action by the Government to halt the slaughter. They suggested that dog licences should be made dearer and that every owner should be required to carry third-party insurance.

Stolen: 15,000 Majestics, colour brown

Daily Mail *Reporter 14.4.62*

The loot snatched in yesterday's big theft was without glitter or sparkle. It weighed 2½ tons and was in 50 heavy brown sacks. The haul: more than 15,000 potatoes – enough to feed 600 families for a week.

Last night, police issued an understandably vague description of the stolen property – "Majestics. Average size. Colour, brown. No distinguishing marks on skins." Not that there is much hope of recovering the potatoes. Police are sure the job was done by an organised gang and that the Majestics are already on the way to the Black Market, on which they will fetch high prices in the present potato shortage.

They were stolen from a Manchester warehouse. A director of the firm, Mrs Olive Pollitt, said last night: "Their wholesale value was £105."

These animal drugs do no harm

John Winter 18.5.62

There is no evidence that feeding antibiotics to farm animals at the permitted amounts does any harm to the animals or to human health.

After slaughter there are no more than traces of antibiotics in the carcasses or in the animal products. These are unlikely to have any ill effects on human health, says a report by a joint committee set up by the Agricultural Research Council in March 1960. The committee recommends that the present permission for the sale of the antibiotics penicillin, chlortetracycline and oxytetracycline in, or for addition to, feeding stuffs for pigs or poultry, other than breeding stock, should be continued. In addition, the use of these antibiotics in feeding stuffs for young calves should be allowed.

It does not recommend any relaxation of the present regulations that would sanction the feeding of antibiotics to adult livestock, including laying poultry, because of the greater risk associated with prolonged use of antibiotics.

Farm price   *John Winter 1.6.62*

£191 an acre was paid for an 844-acre farm at a Cirencester, Gloucestershire, auction yesterday.

Royal Show on the move
John Winter 19.6.62

The Royal Show at Newcastle-on-Tyne may be moved next year to Stoneleigh, Warwickshire for at least five years.

Two gales damaged pavilions during preparation for this year's show.

The 1962 Royal, to be held at the beginning of July, will be run at a loss unless attendances near the record 242,548 of 1956 – when a gale struck and the show lost £13,000.

New-look Royal to put over new ideas   *John Winter 6.7.62*

Getting new agricultural techniques over to the farmer will be one of the main aims of the Royal Show when it opens on its permanent site at Stoneleigh in Warwickshire next year.

Mr Francis Pemberton, who will be the show's honorary director next year, said last night: "There will be a great development in staging practical demonstrations to get new techniques over. If we can do that and at the same time put on a big spectacle, so much the better. We want to give the show as new a look as we can."

Yesterday more than 50,000 braved a winter-in-July day to attend the third day of this year's show at Newcastle.

Daisy the duck loves the patter of tiny, webbed feet
John Winter 31.8.62

Probably no one has ever oozed so much mother love in so short a time as Daisy the Duck. Daisy shares the matrimonial poultry house of Donald the Drake with another duck named Daffodil. And since June any egg that either Mrs Duck has laid homely Mother Daisy has tucked away under her feathers. Twenty-seven in all she collared and sat on, in a breeding marathon unequalled in Duck-land.

Night after night, like little quacking machines, Daisy and Daffodil produced large, fertile eggs. Owner Bert Hall, a retired roadman, of Moorland, Somerset, took some away and sold them. Daisy grabbed the others. For ten long weeks, while Donald and Daffodil were out waddling, she sat. She brought 24 ducklings into the world. Some of them were strapping, eight-week-old adolescents when she helped the last fluffy batch out of their shells four days ago.

Mr Benjamin Adams, a friend of Mr Hall's, said yesterday: "The oldest are so big we shall be having them for dinner soon. The youngest are just tiny handfuls of yellow down."

A Ministry of Agriculture expert said: "A freak occurrence. It is so contrary to the reproducing habits of the duck, it is hard to believe. I have never heard of anything like it before."

1963

This was a relatively quiet, pre-election year that saw the political parties set out their policies for the industry. On the food front prices were seen to rise in the spring after one of the coldest and longest winters on record. There was a threat to the smaller rural slaughter houses and the general drift from the land by both farmers and workers continued. Concern was starting to be expressed about the welfare of animals in transit; but on a lighter note it was the year "that chickens wore glasses"!

 POLITICS and FOOD

Lament for farmers

John Winter 9.1.63

To be a farmer's boy no longer implies a settled future of ploughing, sowing, reaping and mowing. An agricultural economist said yesterday that thousands of farm families would have to leave the land if agricultural prosperity is to be maintained.

Professors are often wrong but we have an idea this one is right. The drift from the land, which has continued for decades, stimulated mechanisation. This, in turn, speeded the drift from the land and made British agriculture among the most productive in the world.

Machines are insatiable. As they gain power and speed they demand more and more space. The small fields of Britain do not satisfy them. For full scope and efficiency they need much wider areas – the prairie technique. This implies the disappearance of the hedgerows, the charming, unique patchwork of the British agriculture, and a cherished way of life. It is a sorrowful thought.

In six years 120,000 workers have left the land. Today the average is two workers to a farmer, and soon the ratio will drop to 1½ to one. Even this is not low enough, says Professor Britton, of Nottingham. Farm population must shrink by 5 per cent a year from the present rate of 3 per cent. This means that many farmers must themselves leave farming, but how this is to be done without undue dislocation or distress is not indicated. Perhaps it will be a natural process.

In that case it will be slow. Men who live by the soil are tenacious of the soil. Their grip will not be quickly or easily loosed. But it is almost inevitable that what is happening in other countries will also happen here.

We should lament it – but we can see no way to stop it.

Dutch Cheddar cheese

John Winter 16.1.63

Shipments of "Cheddar cheese" made in Holland and heavily subsidised will be dumped on our market this year if the Government takes no action to stop it.

Mr Richard Trehane, chairman of the Milk Marketing Board, told a conference of milk producers in London yesterday that Dutch producers, who invaded the British market with 1,000 tons of their brand of Cheddar in the last quarter of 1962, are planning to boost this to 4,000 tons in 1963. This is about 6½ per cent of Cheddar cheese produced in Britain last year and sufficient to upset a market finely balanced between supply and demand.

Mr Trehane said the Dutch had a problem to dispose of surplus milk but, he went on: "I don't think it is sensible to find a solution by indulging in what is a most outrageous form of dumping."

The Dutch bid had followed pressure on the British cream market from Denmark. The board had raised the matter with the Government, but had not got much help. Discussions had been held with the Danes which gave hope of reasonable prices and marketing policies being agreed, and similar discussions were now being sought with the Dutch.

Answering critics of the board's new incentive scheme for quality milk, Mr Trehane said the era when quantity matters more than quality had gone. In a short time the TT grades designation was likely to cease and the new scheme was a step towards giving the consumer a quality alternative.

Who has fresher vegetables?

John Winter 21.2.63

Most housewives accept that vegetables are fresher at greengrocers than at supermarkets, according to a survey published yesterday. They also say they get better value for money and shopping is far more pleasant.

The survey, commissioned by the Horticultural Marketing Council, was into housewives' attitude to vegetables. The report, which the council says is not necessarily the council's view, states that 72 per cent of housewives interviewed thought the greengrocers' vegetables were fresher than the supermarkets'; 59 per cent thought they got better value for money; and 57 per cent said shopping was more pleasant at the greengrocers.

The most widespread criticism of greengrocers (by 77 per cent of those interviewed) was "that they put nice things in front and serve you from the back if you let them." Nearly half (49 per cent) said many greengrocers "fiddled" the price by asking them to take a little more than they wanted. More than 40 per cent said the shops were unnecessarily dirty.

Reasons given by housewives for serving vegetables at meals were: health (64 per cent), filling (10 per cent), flavouring (10 per cent), add interest to meat dishes (7 per cent), variety (7 per cent) and making meals look attractive (2 per cent).

Potatoes were the most popular, followed by tomatoes, peas and lettuce. The percentage consumption was: potatoes 97, tomatoes 93, peas 92, lettuce 86, cucumber 64, beans (any kind) 59, carrots 57, cauliflower 53 and spring onions 53.

The report says 26 per cent of households consumed more than 7oz of vegetables per person per main meal, 61 per cent from 3oz to 7oz, and 13 per cent less than 3oz. Heavy consumption was above average in London and the South-East and below average in the North-West of England and Scotland.

On average housewives served eight to nine different vegetables, seven to nine being the most common number. Fresh vegetables were used by 99 per cent of housewives, tinned by 59 per cent and frozen by 28 per cent.

Liberal Party policy

John Winter 28.2.63

A managed market for agricultural products, with a fair return to our own farmers, co-ordinated with planned imports, is the aim of the Liberal Party's agricultural policy. It considers that rejection of our entry into the Common Market does not mean permanent separation from Europe, and no opportunity should be lost to keep Britain in step with the Six.

Meanwhile, Mr Emlyn Hooson, MP for Montgomeryshire, said at a press conference yesterday, British agriculture faces greater problems than if we were in the Common Market. Outlining the Liberal plan he emphasised that British farming must not be placed on the sacrificial altar, because of the Common Market failure. On two of the major commodities most sensitive to import pressure – meat and cereals – the Liberals would set up commodity commissions, with power to intervene on imports, and to buy in the market to prevent undue price depression.

The policy of the 1947 and 1957 Agriculture Acts had been frustrated by Tory incompetence, which had managed to raise the cost of farm support to the taxpayer and the price of food to the consumer, while, at the same time, depressing farmers' incomes.

Encouragement for improved farm marketing would go hand in hand with the determined attack on farm costs – reasonable long-term credit through a land bank, cheaper machines, fertilisers and other requisites through removal of tariffs and price rings, better capital grants for small farms, and the end of the "death duty racket" by agricultural land speculators and removal of fuel oil tax.

Up go potato prices

John Winter 6.3.63

Potatoes are shooting up in price after the Big Freeze, which has ruined thousands of tons.

Latest rises have not yet reached the shops, but farmers are getting up to 50 per cent more for frost-free supplies than they did at this time last year. And by May last year prices had doubled as Britain headed for the worst potato shortage since the war.

A warning about the precarious balance of potato production came yesterday from Mr James Rennie, chairman of the Potato Marketing Board. He forecast a "disastrous fall" in potato acreage unless the Government raises the standard price. Mr Rennie was speaking at Haddington, East Lothian, eight days before the Minister of Agriculture, Mr Christopher Soames, is expected to announce the results of the February price review. This will list guaranteed prices for the next 12 months for 13 commodities, including potatoes.

An example of the rise in farmers' prices: last week at Chelmsford, Essex, King Edwards brought £27 a ton. Comparable price last year, £22.

Tory Party policy *John Winter 13.3.63*
First outlines of a new Conservative agricultural policy are expected to follow the announcement of the farm price review settlement on Wednesday.

During his month-long talks with the NFU team, Mr Christopher Soames, Minister of Agriculture, has sketched the Government's ideas for limiting capital Exchequer liability for farm subsidies, and tightening control of food imports by tariffs. The Government's aim is a half-way house between its present support system and the complete elimination of subsidies demanded if Britain had joined the Common Market.

However this is achieved it will raise the cost of the housewife's food bill.

Quantitative limitation of price guarantees, already applied to milk and potatoes, are likely to be extended to other products. However, others, which are subject to serious import competition, may require some form of Government agency with control over imports and power to bolster the market at times of surplus.

To create the best atmosphere for winning farmers' confidence in its future programme, the Government has striven to achieve agreement for the review settlement itself. Price cuts which are regarded as almost inevitable on eggs, pigs and barley, may be softened by increases on one or two other items. The Government could also make some allowance for losses farmers have suffered through the weather. On review commodities alone, these losses are estimated at £3 million.

Revolt against Price Review

John Winter 16.3.63
First rumbles of a farmers' revolt against the price review settlement came yesterday, less than 48 hours after it was announced.

Wiltshire farmers sent a telegram to Mr Daniel Awdrey, Tory MP for Chippenham, saying they had no confidence in the Government. They urged him to "oppose this continued burden placed upon the agricultural industry." In Somerset the NFU described the settlement as a harsh one which would undermine farmers' living standards.

At the annual luncheon of the Devon Long Wool Sheep Breeders' Association at Okehampton the chairman, Mr R W Darke, deplored the results of the review. He pointed out that 39 per cent of the farmers in the country had earned £600 a year less in the past 12 months. He was unable to see how the extra ½d a gallon on milk, resulting in £30 extra income for the small producer, could possibly help in the present situation.

Rates cut will save 6*d*-in-slot milk machines *John Winter 27.4.63*

Sixpenny milk machines have had a reprieve. It looked as if they would disappear because of the heavy rates being slapped on them. But now the Lands Tribunal has decided they should be rated only on their site value, instead of their cost, which is about £365 each. This is great news for machine owners, makers and dairymen. And, of course, for the Milk Marketing Board, which says the machines have sold 248 million extra pints in five years.

The march of the mechanical cows has been impressive. In 1957 there were four. Last year there were 6,000. But rating assessments soared from a few pounds a machine to as high as £40 and £50 and the yearly increase of

1,500–1,700 new machines dropped to a mere 550 last year.

Only 49 new machines have been set up in the first three months of this year and the Milk Board feared that hundreds of them would soon be taken out of service. High rates were largely responsible for this. But machine owners took another knock last September when milk went up by ½d a pint. They could not charge the extra ¼d for the half-pint cartons, and were forbidden to sell a smaller amount for 6d. As each machine has to sell 90 cartons of milk a day before it breaks even many were not making any profit at all.

The board is trying to get the Weights and Measures Bill, which is now before Parliament, amended. The amendment would allow machine owners to sell milk in third of a pint cartons as soon as the Bill becomes law instead of after two years, as is now proposed.

As you tuck into your pork today
John Winter 3.6.63

More and more housewives are buying pork for the family dinner, says the Pig Industry Development Authority's quarterly intelligence summary.

Pork has become such a popular dish that we are now eating 21.6lb a head a year, which is almost double the pre-war average of 10.6lb.

In fact, pork is rapidly catching up with mutton and lamb in the popularity stakes. Before the war we each ate 25.2lb, but now the figure is down slightly to 24.7. Even the traditional beef of old England has lost ground from 54.9lb to 51.4lb.

Pig farmers will also be heartened by PIDA's news that pork imports fell to 2,000 tons in the first three months of this year, or little more than a third of the amount imported in the first quarter of 1962. Home production this year between January and March was 128,000 tons, a 4 per cent increase over 1962.

On a long-term but still heartening note, the PIDA report heralds a slight dropping off in sow numbers as a possible beginning to a decline in the size of the national pig herd. It remains to be seen how much effect the Government's re-enforced flexible price-guarantee arrangements will have on the levelling out of pig numbers.

Pork pig prices at auction markets have fallen off in the first quarter of this year from 45s to 35s a score deadweight, and there was a similar drop in prices for cutters and baconers. Over the same period feed costs have hardened. Maize meal went up by more than £2 a ton to £26 10s, and barley from £20 to £20 17s a ton.

Farmer Brown's pledges 1s
John Winter 18.7.63

Mr George Brown, deputy Labour Party leader, last night promised housewives that their food bills would not be higher under a Labour Government.

He promised taxpayers that their obligations for home-produced food subsidies would cease to be "open-ended" or unlimited.

And he promised farmers most of what they have recently been demanding from the present Government – a continuance of their existing guaranteed price system, control of food imports and more producer marketing boards. Plus a world food plan to channel surpluses from countries of over-production to the starving populations of under-developed areas.

His promise that food prices would not rise came after he had admitted that the world food plan, and some of the other proposals in Labour's agricultural policy, were bound to cost money. Mr Brown said: "If you want stability for the home farmer you have got to have it for overseas farmers as well. You have got to get food into the mouths of starving Asians, Africans and Arabs. If it costs us a bit of money, it will still pay us. I don't believe it will put prices up. It will bring stability to the industry. It will control the consumers' problem and limit the taxpayers' open end."

Mr Brown was speaking with Mr Fred Peart, Labour Shadow Cabinet Minister of Agriculture, at a Press conference in Norwich before addressing a mass meeting of farmers, farm workers and consumers at Swaffham on details of a 14-point policy for agriculture under a Labour Government. Claiming that all good features of the Government's agricultural policy stemmed from a 1947 Labour Government Act, Mr Brown said that the Socialists, unlike the Tories, now considered that agricultural problems must be viewed and settled on a world basis.

Slot machines   *John Winter 9.8.63*

Crippling rates and rising costs threatened to wipe the 6d-in-the-slot milk machines off the streets of Britain earlier this year. That is all over now. The Milk Marketing Board joyfully predicted yesterday that the 6,100 machines, which have sold an extra 250 million pints of milk in five years, will be increased to 7,100 in a year's time.

Easing of the rating burden and Government legislation to allow machine operators to sell slightly less than half-pint cartons for 6d, means vending machines are an economic investment again. Holiday campaigns at South Coast resorts have sold 14,000 extra pints in a matter of days. Machines at Brighton dispensed 10,000 half pints in three weeks and a mobile milk bar at Sandbanks, Dorset, sold 10,000 in eight days.

Weekend food   *John Winter 8.10.63*

Most eggs will cost more this weekend. Only standard Lion eggs are unchanged at 4s to 4s 3d a dozen. The others are all up by 3d a dozen with large at 5s 3d to 5s 6d; medium 3s 3d to 3s 6d; small 2s 9d to 3s.

Bacon prices remain steady.

Butter and *cheese*: no change.

Fruit: Cheaper alternative to Cox's apples now costing from 1s 3d to 1s 9d are Ellisons Orange, Charles Ross and James Grieve varieties at 10d to 1s. Bramleys are 1s 3d; Conference pears 1s 6d to 1s 9d; Almeria grapes 2s to 2s 6d; small grapefruit 8d each; melons 2s to 3s.

Vegetables: Carrots cheap at 4d to 5d a lb; cabbage 4d to 6d; sprouts 8d to 1s; cauliflowers 10d to 1s 2d each; celery 6d to 1s a stick; leeks 1s a lb and watercress 6d to 8d a bunch. Parsnips, swedes and turnips 5d to 6d a lb.

Meat: Little change in price except for slightly cheaper cuts of New Zealand lamb. Recommended buy: English shoulder of lamb at 2s 6d to 2s 8d a lb.

Fish: Codling fillet 2s 9d to 3s 6d; coley fillet is 1s 8d to 2s; whiting 1s 2d to 1s 6d; plaice 2s to 4s; herring 1s 4d to 1s 8d.

Sugar prices should stay around the 1s 7d for 2lb mark.

One-man campaign

John Winter 21.10.63

Farmer Wallace Day, well known for his revolutionary views, launches a one-man campaign today for a better deal for farmers, and greater security for British agriculture.

In intervals in the struggle to gather his storm-battered harvest Mr Day, 47, chairman last year of Devon NFU branch, has written a 25,000-word study of the industry's political problems, published today as a 64-page booklet. It advocates redistribution of subsidies, levies on food imports, and slightly higher prices in the food shops, to enable farmers, especially small ones, to keep pace in income with the rest of the community.

His plan for a complete revision of the Government price-support programme over the next five years would cause a big row, because he says the subsidies of wealthy cereal farmers should be cut to increase those of small men who depend on livestock. He claims: "Large corn-growing farms are receiving far more than their share of the farming budget. The recipients of this easy money are responsible for the 'rich farmer' legend."

Mr Day condemns the booming "barley beef" craze, dependent on three subsidies, and encouraged by present Government policy, as a distortion of farming patterns, which will do irreparable harm to the industry. Beef cattle will be driven off the land, because grass feeding cannot compete with this three-line artificial husbandry.

Potatoes, he believes, should be excluded from price guarantee, because growers can pay for their own support through the Potato Marketing Board. For last year's crop 18,000 of them had more than £15 million above the guaranteed price. This inflated income, taken into account in the price review, depressed the prospects of all farmers, although only one in four grows potatoes.

The housewife, too, would feel the effect of Mr Day's plan. He would like to see the 5s 6d of every £ of personal expenditure, which at present goes on food, raised to 6s. Farmers get only 1s 2d in the £, plus the equivalent of 4d in grants and subsidies. The extra 6d would remove the need for subsidies and give them fair prices.

The booklet, *Fair Prospect for Farming*, costs 7s 6d.

Threat to slaughter houses

John Winter 2.12.63

The Government is considering a plan to "do a Beeching" on the meat industry by closing hundreds of slaughter houses and concentrating preparation of home-produced fresh meat into a few huge slaughter centres. This would be part of a programme to create a market with regulated supplies of both home and overseas meat, a key feature of the new food policy of Mr Christopher Soames, Minister of Agriculture.

The plan is backed by the National Farmers' Union and an influential group of agricultural MPs.

Agreements to fix the volume of beef and lamb to be sent to Britain are being negotiated with overseas suppliers and these will stipulate the rate at which the meat is shipped. But if Mr Soames is to achieve his aim of market stability, by ironing out shortages and surpluses, he faces the more complicated problem of controlling the flow of meat from home farms. He is awaiting the report of a committee of inquiry into meat marketing set up last year and headed by Sir Reginald Verdon-Smith.

Also important is the avoidance of collisions between imports and home-produced meats. Early this year heavy shipments from Argentina clashed with unexpectedly high marketing of cattle from our own farms. Prices slumped and subsidy payments on home production soared. The slaughtering stage is being recommended as the logical regulation.

The Verdon-Smith committee is expected to come down against a producer-controlled marketing board to buy and sell all home meat, which the NFU has proposed. Instead it is likely to recommend a purely regulatory marketing commission to administer both home and overseas supplies.

CROPS and MACHINERY

Chemical weed control

John Winter 2.3.63

Weed control in potatoes without mechanical cultivations may be possible by combining the use of chemical weed killers and smothering the weeds by closer planting of the crop, suggested Mr R G Hughes, regional crop husbandry officer, at a Cobham, Kent, meeting.

First-class grass   *John Winter 13.3.63*

After scouring the world for grasses to extend the grazing season, scientists at the Grassland Research Institute, Hurley, Berkshire, have found one almost on their own doorstep.

It is S.170, a tall fescue, which has been on the market for 15 years and originated from plants found about 20 miles from Hurley, near Aylesbury, Buckinghamshire. S.170 gives good grazing into mid-December and provides an "early bite" two or three weeks before most other perennial grasses in spring. Its hardiness also beats foreign competitors. It came through the severe winter almost unscathed, while North African tall fescues on trial at the institute were set back severely.

Commercial use of S.170 on 30 farms in England and Wales has confirmed the hopes of the institute's scientists. Dr Kenneth Baker, head of the extra-mural department, says S.170 could be a boon to small dairies and other intensive livestock farmers. If rested for six weeks from the end of August, it produces about one ton of dry matter an acre with digestibility of 72 to 73 per cent. Dr Baker says that in spring it is as early as a 12-month-old Italian ryegrass and cattle thrive on it.

To get best results it needs highly fertile soil, good management and a high stocking rate of about 1 to 1¼ cows to the acre.

Working overtime

John Winter 23.3.63

Night ploughing and drilling by car and tractor headlights is spreading in Norfolk in the effort to make up for the months lost in the cold spell.

Big freeze affects watercress

John Winter 26.3.63

Watercress beds took such a hammering in the Big Freeze that it will be months before housewives see plentiful supplies at normal prices. Only a trickle of bunches is coming off the 100-odd watercress farms, and small supplies reaching the shops are about twice the usual price. Small bunches cost 9*d* to 1*s* each.

Growers have never known a worse early season. From beds which should be in peak production, some have dragged tons of brown, rotting cress. Electric power cuts aggravated the position on some holdings by putting out of action the pumps which drive water through the crops. The beds became solid ice.

Beds are now being re-seeded, but the first new crops cannot be expected before May. Full production may not be reached until September. Losses have been high. One Hampshire grower estimates his at £10,000. Another in Dorset says he is sending less than a quarter of his normal loadings to market.

More bulbs than ever before

John Winter 15.4.63

Record acreage of flower bulbs is being grown in North Lincolnshire this year.

New technology for blueberry pickers
John Winter 18.4.63

Latest cost-cutting ideas in American mechanised farming: a power "drill" which picks blueberries, an electronically controlled asparagus cutter, and a wind machine for harvesting grapefruit. They are being developed at Michigan State University and research centres to meet farmers' problems caused by an uncertain labour supply and rising wages.

The electric eye on the asparagus cutter singles out stalks six inches or more, leaving shorter ones to ripen. The drills, four to a machine, shake the plants, causing ripe berries to fall into a canvas catcher. The wind machine has been devised to simulate hurricanes, which strip orchards of fruit. It is estimated that it will harvest 150 trees in an hour, a three-week job by hand. But a prototype, which removed 99 per cent of the fruit, shredded all the leaves as well!

Flower growers beat the weather
John Winter 18.5.63

Flower growers of Spalding, Lincolnshire, show the world today how they beat one of the worst winters on record. Despite the Siberian weather and the late, cold spring they are staging their flower parade as usual. It is two weeks late, but better late than never. For they have had to use later varieties of tulip, which give a much wider range of colours.

Since early yesterday, and all through the night, about 300 of the 12,000 people who work in the Spalding flower industry have been preparing the tulip-decked floats. They have had to pin about eight million tulip heads on to the floats, some of which are 40ft long by 12ft high and cost up to £1,000 each.

But it is money well spent. The mile-long parade of floats and decorated vehicles is a marvellous shop-window for these South Lincolnshire growers who sell ten million boxes of flowers each season, and for the British flower industry. This year 200,000 people are expected to watch the parade. Hundreds come from abroad specially to see it.

The theme of this year's parade is Worlds of Flowers with a different world depicted in a blaze of colour on each float.

But the designers have had to face the problem which has defeated them for years. There are no blue tulips. Organising secretary Mr C J Vivian said this made sea or sky scenes very hard to do. "Still, we will get a blue tulip one day. We have got breeders working on it."

One float is sure to raise a cheer this year. It is a moon-probe rocket, plus the moon with American and Russian flags.

No trade for early potatoes

John Winter 12.7.63

Early potato growers have been told to leave the potatoes in the ground because it's not worth marketing them.

This is the advice given by the Potato Marketing Board yesterday after holding an emergency meeting to consider the big slump in prices. Home-grown potatoes are selling for as little as 3*d* a lb, which means growers are getting less than 1*d* a lb.

If necessary the board is prepared to use its powers to buy potatoes from registered producers from August 1. In the meantime it hopes its warning will cut supplies and increase prices on wholesale markets. The board is also increasing the riddle size by half an inch to one and a half inches from Sunday, which means that growers will be able to sell only better quality, larger potatoes.

The depression in the potato market has been caused by several reasons, including bad weather, heavy imports and a late season in which earlies from different areas have come on the market at the same time. Clearance of the early acreage has also been slower than in the past two seasons. By the end of this week about 30,000 acres will have been lifted compared with 37,500 last year and 50,000 the year before.

This means that more than normal quantities are coming on to the market later and are threatening to depress the prices for the first main crop potatoes.

Main crop acreage is 30,000 acres more than last year but prospects are uncertain because blight is reported in the main growing districts. This could reduce yields considerably.

Easing the load   *John Winter 1.8.63*

Bulk handling of grain and feed to and from British farms is steadily rising, says a booklet published today by the National Association of Corn and Agricultural Merchants. It says that between 30 and 40 per cent of the wheat and barley moved from farms in the main growing areas and about 20 per cent of poultry food are bulk handled.

Barley price slump threatens subsidy bill   *John Winter 2.8.63*

Measures are expected today to prevent cheap imported grain from knocking the bottom out of the home barley market before it has opened.

Urgent demands for Government action have poured into NFU offices, especially from the cornlands of East Anglia, since it became known that French exporters were offering barley for forward delivery at as little as 17*s* a cwt landed price. Even allowing for the 10 per cent import duty, this easily undercuts the 19*s* 6*d* a cwt guide price for the home crop, a price which was recommended three weeks ago by the working party representing all sides of the trade. Under the threat the guide price has already been cut to 18*s* 9*d*.

The crisis comes just as the heat wave has given farmers an unexpectedly early start on harvesting a barley crop estimated to reach a record six million tons. A price slump would run up a subsidy bill far above the estimated £36 million, just when Mr Christopher Soames,

Minister of Agriculture, is negotiating with home and overseas producers to put a limit on the "open-ended" subsidy.

Brighter outlook for the harvest

John Winter 10.9.63

The harvest could be a lot brighter than many farmers expect, provided we get normal weather for a few weeks. This ray of hope came last night when the Ministry of Agriculture released its official countrywide survey of crop prospects.

All cereals were battered by high winds and heavy rain during August. But the report added: "Yields were expected (at August 31) to be slightly above average, given good normal conditions for the rest of the season."

Although there has been more than a week of almost continuous bad weather since, the Ministry estimates of yield still give a good basis for hope. The estimates are all higher than those made at the end of August 1962 for last year's crops. Then the weather had been similar until the last week of the month, which was warm and dry.

Yield estimates in cwt per acre:

	1963	5-year average	1962 estimate	1962 final yield
Barley	26.5	25.8	25.5	29
Wheat	30	28.4	29	32.2
Oats	23.5	22.4	23	25.2
Mixed corn	23.5	22.4	23	25

Little harvesting has taken place, the Ministry report continues. There was fairly widespread take-all and eye-spot diseases in wheat and lodging in most areas but crops were standing well. From all parts of the country there were reports of laid crops of barley with second growth and much green corn, and the harvest was expected to be difficult and protracted.

Sugar beet was growing well, but needed sun to promote sugar content and yield, which should still be above average. Estimated yield for maincrop potatoes was 9.1 tons an acre compared with the five-year average of nine tons.

The hay harvest had been slow and laborious but yields were above average, although quality was poor.

The NFU yesterday estimated that damage to cereals could represent a loss of £40 million.

 # LIVESTOCK and POULTRY

Lambing by lamplight

John Winter 1.1.63

Electric light is to be tried on sheep as a means of extending the spring lambing season and adding a lambing in autumn. The lights will be switched on to flocks of sheep to see whether they affect lambing.

The trials will be held on a Berkshire farm early this month by the Electrical Development Association and the Department of Agriculture at Reading University. Object of the experiment is to transmit to other breeds the characteristics of the Dorset Horn, which not only starts lambing in the late autumn or early winter, but sometimes produces two crops of lambs a year. This was long attributed to temperature and condition of pastures, but it is now thought that light is a more dominant factor.

Lamps of 1,000 or 1,500 watts will be used for the flash treatment. One flock will get the equivalent of an extra five hours' daylight while a second flock will have an extra five hours' darkness. Promoters of the trials are confident that with high standards of feeding and management the sheep will come to no harm.

If the trials are successful widespread adoption of the technique could radically change the pattern of sheep husbandry and put the production of fresh home-killed lamb on an all-the-year-round basis.

The trouble with tails and why a little piggy has none

John Winter 21.1.63

The piglet which bites off the tail of its fellow piglet is worrying pig farmers.

No one knows why they have started to nibble each other's tails. So a feeding stuffs firm, British Oil and Cake Mills, has invited thousands of farmers to join in an investigation. It has sent out a questionnaire seeking facts about the "vice", and the methods used to check it. Replies will be analysed by electronic computer to seek out common factors. From these it is hoped to establish causes and work out a cure.

Bitten tails lead to loss of growth and even death. Damage is permanent, and a tail-less pig produces an imperfect carcass. Mr David Bellis, the firm's chief pig adviser, said: "Tail biting has increased as pig rearing has become more intensified. In some cases it has grown so bad that big pig producers are detailing their pigs soon after birth. Unless we can stop it we shall get a population of 'Manx' pigs."

Tail biting is likely to start when piglets are a few days old, and some farmers blame boredom. Ideas for checking it include keeping pigs in darkness, except at feeding times, and scattering salt on the pen floor.

Effective vaccine

John Winter 29.1.63

Vaccine against fowl pest has saved large poultry flocks in Norfolk during the recent outbreaks. In one case nearly 20,000 poultry and 10,000 turkeys owned by Mr M Grief of Shrublands Farm, Northrepps, survived a severe attack on the premises after two doses of the subsidised vaccine.

Evulse goes for £63,000

John Winter 6.2.63

Lindertis Evulse, a 14-month-old Aberdeen Angus bull yesterday became the most valuable chunk of beef in the world. Ranchers and cattlemen from all over the world battled with bids at a Perth auction to own him. Every nod they made was worth 1,000 guineas.

After four frantic minutes, the top bid of £63,000 was made on behalf of Mr Jack Dick, managing partner of a cattle breeding syndicate in New York State. That makes Evulse worth £70 per lb, and "well worth it", said Mr Dick. "He is a bull that can't be faulted." He added: "The price we paid is probably not as far as we would have gone. We consider he is good value at that price, but we would, if required, have put up the bidding."

The previous world record for a beef animal was £47,000, which was paid in Argentina.

Evulse's brother

John Winter 7.2.63

Half-brother of Lindertis Evulse, the Aberdeen Angus bull sold for a world-record £63,000 at Perth on Tuesday, fetched only £75 at the sale.

Beef from British bulls

John Winter 1.3.63

Herds of young bulls will be frisking around British farms if tests to be started this year are successful. The National Farmers' Union, backed by the Ministry of Agriculture, has decided to experiment with bull-rearing for beef.

The Ministry has agreed to permit the rearing of 500 bulls to the age of 15 months for slaughter for beef so that their carcasses can be compared with those of bullocks of the same age and breeding raised side by side with them. This departure from the beef-rearing routine followed since the war has been precipitated by the success in British butchers' shops of imported Yugoslav beef, produced from bulls.

Until now, bull-rearing for beef, beyond the age of ten months, has been illegal and completely uneconomic. It has been accepted that castrated male animals – bullocks or steers – produce the right sort of beef, and they are also less dangerous around the farm. Ministry regulations insisted that any entire bull kept beyond the ten months must be licensed, and only bulls which would make suitable breeding sires were accepted. Calf subsidy and fatstock guarantee were only paid on bullocks and heifers. Now farmers believe that young bulls, to be slaughtered at 12–15 months, put on weight more quickly and economically, and also produce leaner and more tender beef – the kind modern housewives prefer.

Licensing and subsidy rules have been relaxed by the Ministry for the tests, and the NFU is to nominate farmers who will agree to rear bulls and bullocks on a comparative basis. In England and Wales, 25 farmers will each rear ten bulls, while 50 more bulls will be reared on Ministry experimental farms, some to a greater age than 15 months, to find whether the advantages claimed continue in older animals. Scotland will raise 150 bulls, and Northern Ireland 50.

Bringing swine fever under control

John Winter 20.3.63

The Government's new scheme to conquer swine fever by compulsory slaughter and compensation has got off to a sky-high start. Its first week will cost the taxpayer £350,000. This is the bill for compensating the owners of 27,000 pigs slaughtered since the scheme started on March 11.

The 300-strong veterinary staff of the Ministry of Agriculture have been working almost round the clock. They were already busy checking poultry flocks for fowl pest, for which slaughter and compensation continues until the end of March, when they moved into the swine fever scheme at the height of a major epidemic. Over 150 herds in 34 counties were already suffering from the disease. These were immediately scheduled for slaughter and another 78 herds have since been struck down.

Only a tiny proportion of the slaughtered pigs were actually showing symptoms, and they qualified for compensation at half their value. The others, which had to be slaughtered as contacts, will be paid for at full value. All animals, whether diseased or not, are destroyed and there is no salvage.

A Ministry spokesman said: "In the circumstances in which it started the new scheme involved a huge volume of extra work, but it is operating smoothly." The number of outbreaks and slaughterings is expected to fall steadily.

Despite its expensive start, the

Ministry is hopeful that the scheme will soon bring under control a disease which averaged 1,000 outbreaks a year up to 1961 and reached nearly 1,900 last year. A pig expert has estimated it can eradicate swine fever in two years.

Soundproof home for Superpig

John Winter 19.8.63

Baby superpigs will be scuttling around their aseptic quarters, breathing filtered air and drinking their pasteurised milk substitute down on the bug-free farm this week.

But if you go, you will not see a single curly tail, nor hear the faintest squeak. The building, which houses the first batch of mini-supers at the new minimal disease farm of FMC (Meat), at Calne, Wiltshire, is sight-and-sound-proof. For their 130-mile journey from the laboratory at Cambridge, where they were born by surgical methods, the mini-supers travel in incubator-like compartments, in a van fitted with a filtered air-conditioning plant.

Mr Lewis Jollans, a 37-year-old geneticist in charge of the farm, said: "One attendant looks after the food and the cleaning, and we only peep at them occasionally to make sure they are all right. They look fine and we are confident that the precautions taken at birth and the routine observed here will rule out internal and external parasites, all bacteria and nearly all viruses."

The scientists visited farms from Scotland to Cornwall to collect sows of impeccable breeding, mated to equally well-bred boars. The present intake of 128 piglets will be repeated eight times at ten-week intervals. Half of them will be Large Whites and half Landrace. The super-stock will be created by crossing the two breeds.

The day a chicken saw red

John Winter 2.10.63

A seven-month-old white chicken was laid flat on a table in a Lytham St. Annes hotel yesterday, and held by two scientists while rose-coloured contact lenses were inserted between the lids and corneas of its eyes. When it was released, it scratched at its head, but stood completely still – because its vision was distorted and it "saw red".

Contact lenses for egg-layers were invented in the United States to make the birds placid, and stop them pecking each other, spilling their food and breaking eggs in multiple batteries, where up to 20 birds live in one cage. So far they have been used only experimentally in Britain, at a research farm at Stoke Mandeville, Buckinghamshire.

Tests there are directed by Dr William Blount, leading poultry scientist, who demonstrated the contacts yesterday. Plastic discs of standard size are used for any kind of chicken, he explained. They can be inserted into the eye when the bird is six to eight weeks old, and left there for life. The length of time they remain in position is not yet known. Dust or germs, introduced at the time of insertion, might cause inflammation.

The lenses do not magnify, but their shape, with a tiny bulge in the centre, causes distorted vision, while the red colour has a tranquillising effect and prevents the bird from seeing blood, even if it is shed.

"At Stoke Mandeville we have found no ill-effects from contacts, and the birds fitted with them showed no signs of discomfort," said Dr Blount. "There is sometimes an immediate reaction of head-shaking, or scratching at the eye, which quickly passes. For two days the birds ate only half their normal food, but then returned to normal. So far we have tried out contacts only to test reactions, and have not published any findings."

Did he expect any opposition from animal protectionists? – "That is quite possible," he admitted.

Contacts cost 3d a pair, and two extra eggs would cover the purchase and cost of insertion.

How long will the housewife try a little tenderness?

John Winter 15.10.63

The roast beef of our new England is growing leaner, and more tender, but less tasty. And the butchers don't like it. After months of misgivings in cattle markets, abattoirs and the pubs around Smithfield Market, their fears came out loud yesterday.

The new president of the National Federation of Meat Traders' Association conference at Harrogate, Mr Tom Tasker, had these blunt words to say about barley beef, the new, quick-money type of meat production which is spreading through Britain. "This beef is quicker maturing, which means that it is leaner. It is eaten younger, and is therefore tender. But some of this is achieved at the expense of flavour."

Part of the enjoyment of eating beef was the flavour, and that was something which could not be achieved artificially. It came with maturity. If the barley beef craze goes on, Mr Tasker fears that both farmers and butchers will be in trouble. They will be trying to sell an article which has lost some of its appeal.

Mr Tasker, 47, is in a unique position to judge. He is both butcher and farmer. He combines both in partnership with his cousin. On their 500 acres at Tetney, near Grimsby, they produce pigs, sheep and beef cattle. As a farmer, he understands the reasons why beef cattle are disappearing from the pastures. As a butcher he is alarmed and, I think, with good reason.

The reasons for the change are economic. Beef raised on the old pattern is financially the least attractive enterprise in livestock farming. Not only are profits unpredictable, and often small, the length of time in which a farmer's capital is walking about on four legs is too long. An old-fashioned beef beast is two to three years old before it is slaughtered. And it is this which gives the meat flavour. It may also make it tougher, and add more fat to the joints than modern housewives like to see. But nature has made fat and flavour inseparable.

The barley beef barons have gambled that if a housewife sees much lean and little fat in a joint and it remains tender, no matter how it is cooked, she will never notice that it lacks flavour. So they take a batch of calves, usually sired by a beef bull out of a dairy cow, and confine them in a yard. Their indolent life tends to make them put on flesh more quickly, but it doesn't affect their appetites.

At 11 weeks they can be switched to an ad lib diet of barley, on which they put on astonishing weight gains of nearly 3lb a day. By 11 months, or less, they weigh about 900lb and are big enough to slaughter. Each animal which costs £14 as a calf and about £40 to feed with barley, will give the farmer a profit of about £23 when sold.

Attractive money. Yet the main attraction is its speed. And that is also the danger. The temptation will be to make it even quicker, and the earlier beef is slaughtered the more tasteless it will be.

Remember what happened with broiler chickens? At first their life span was 13 weeks. Now it is nine and Mr Tasker has no illusions about the risk to beef. He asks: "Supposing, in a few years, we found that the British family had lost its taste for beef, that we had run into the troubles that faced the broiler industry, with a large section of the community searching for 'free range' food . . .?"

It's a question worth pondering.

 # EMPLOYMENT

Too many accidents

John Winter 30.1.63

The growing number of farm accidents in Essex, Suffolk and Norfolk is worrying the Ministry of Agriculture. Last year 23 people were killed and 2,400 injured.

Farmers go back to school

John Winter 22.5.63

Open-air classrooms for farmers and housewives will provide lessons today at one of Britain's busiest and most enterprising farming exhibitions.

The two-day Staffordshire county show, which opens at Weston-on-Trent, near Stafford, will start up a new demonstration department. The sheep demonstration is the show's contribution to increasing know-how among the county's farmers, who are shaping their businesses to meet keener competition and higher standards in agricultural marketing. Forced by economic pressure to concentrate on fewer lines of production, they are abandoning their traditional pattern of mixed farming.

In the last seven years 2,000 Staffordshire farmers have gone out of milk. In the same period, the sheep population has rocketed. From about 200,000 before the war, it had fallen by more than half after the losses of the severe 1947 winter. Now it is at a record level of 210,000, and farmers are eager to learn the latest methods of economising in labour and feeding costs.

Among the sheep "walks" and dipping "slides" is an ingenious pen invented by a local farmer for holding animals while they are dosed with medicine. At a pull on a lever the false floor rises a foot from the ground and three sheep remain firmly suspended while they are treated.

Sugar beet transport

John Winter 23.5.63

The NFU have given full support for a new road safety by-law in Kesteven, Lincolnshire, which makes it an offence to carry sugar beet in unsuitable vehicles.

We work harder but earn less

John Winter 27.6.63

The farmworkers' case for a minimum weekly wage of £10 and a 40-hour week was argued yesterday at a meeting of the Agricultural Wages Board. The meeting was adjourned until July 31 when the employers' reply will be given. The present minimum of £9 3s came into operation last November and an application to reduce the working week from 46 hours to 40 was rejected in January.

The case yesterday was put by Mr B Hazell, National Union of Agricultural Workers, and Mr T Healy, Transport and General Workers' Union. The employees' case was that the farmworker on average puts in four hours more work a week than the industrial worker and takes home about £4 13s less.

Call for £10 minimum

John Winter 22.7.63

1,000 trade unionists at a mass demonstration at Tolpuddle, Dorset, yesterday pledged firm support for their union's claim for a £10 minimum wage for a 40-hour week which is now before the Agricultural Wages Board. The demonstration marked the 129th anniversary of the Tolpuddle Martyrs, the six farm labourers sentenced to transportation for their part in forming a union, forerunner of the modern trade union movement.

Pay errors

John Winter 5.8.63

The wage packets of 28,144 farm workers were checked by Ministry of Agriculture inspectors last year. They found that 449 workers, 4 per cent of the total farm labour force, on 297 holdings had not been paid the full rate for the hours worked. Wage arrears totalled £12,015.

The present minimum wage is £9 3s for a 46-hour week, but farm workers have put in a claim for £10 for a 40-hour week. The Agricultural Wages Board meets to give its decision on September 11.

 # PEOPLE in the NEWS

Guild appointments

John Winter 5.3.63

Percy Izzard, retired agricultural correspondent and gardening writer of the *Daily Mail*, was installed as president of the Guild of Agricultural Journalists. The guild also welcomed its first woman chairman, Miss Mary Cherry, of *Farmer and Stockbreeder*.

How do you meet the challenge of the changing 60s?

John Winter 28.3.63

A wind of change is blowing through Midlands farming. The farmers are facing up to twin problems: Efficiency and Marketing. They have to produce more food from less land and sell it cheaply in a market which is growing steadily more discriminating.

Take a look round some of these changing farms, and at the men who run them, and you sense revolution in the air. Take Mr Eddie Mander, one of three farming brothers who runs his do-it-yourself farm of 200 acres at Over Whitacre, Coleshill.

The main building in his yard used to be a cinema in Birmingham. It was dismantled and cunningly rebuilt to form a covered yard for beef cattle. Alongside it is a pig-fattening house, built from Nissen huts. Loads of junk are always arriving at the farm, to be quickly adapted and rebuilt into something functional.

As well as running his farm most efficiently, Mr Mander finds time to be a farming politician. He is vice chairman of Warwickshire branch of the National Farmers' Union and vice chairman of the

County Quality Bacon Producers. He told me: "We are tired of buying retail and selling wholesale. It is time we started buying wholesale and selling retail."

He is pressing for central pack houses, where farmers can pool produce and market it in regular, graded lots, cutting out the middleman and doing business with big retail companies direct.

Milk, the biggest single item of farm produce, has become steadily less profitable. But it can still be made a sound business by increasing efficiency and cutting costs. Two farms near the Warwickshire-Staffordshire border show how milk can still be highly profitable.

At Clifton Campville, near Tamworth, Mr Frank Biscoe saw no future in being a jack-of-all-trades farmer. So he got rid of the sheep, pigs, poultry and three other lines of production on his 267 acres and concentrated on corn and milk.

At Edingale, on the Staffordshire County Council farm estate, smallholder Mr Bill Turnock, 36, achieves the same kind of success the harder way. He keeps 50 cattle on 49 acres, and produces 160 gallons of milk a day. Last year his milk cheques exceeded his production costs by £2,500 and this year he expects a bigger balance. Both Mr Turnock and his wife, Norma, work 15 hours a day.

There is even a modest fortune to be made in the chancy business of market gardening, though most growers would query this after the bleak weeks of the Big Freeze. In the Midlands they do this branch of farming the most sensible way, by pooling output and buying. At Pershore, Worcestershire, is Britain's biggest horticultural co-operative – Pershore Growers – suppliers of fruit and vegetables to the markets of Britain. A thousand growers are members. In the past three years its annual throughput has risen from 9,000 tons to 20,000 tons, and its turnover from £416,000 to £880,000.

Long service medal

John Winter 25.5.63

A gold medal for long service will be awarded to 76-year-old Mr Charles Albert Bellamy, still working as a shepherd at Belton, Lincolnshire.

On the move

John Winter 17.7.63

Mr Phillip Sheppy, a 29-year-old organising secretary of Northampton Town and County Federation of Young Farmers' Clubs, goes to a new post with the Royal Counties Agricultural Society.

The Devil and Farmer Rice

John Winter 23.8.63

Time and the Devil have caught up with the Rices, of Morson Farm, after 300 years.

When the auctioneer's hammer falls in a sale at an Okehampton hotel tomorrow, it will close a furrow of farming history, which generations of Rices have been cutting deep into the Devon soil since Stuart times. Time, which has been on their side for centuries, is running out on them. The last of the Rices of Morson, who has decided to sell, is bachelor Mr Francis James Rice.

Exactly how long his family have owned, or tenanted, the 220 acres, which sweep over a gentle hill between the black heights of Dartmoor and Bodmin Moor, is lost in the whispering echoes of local legend, and the dry, fading legal records in a solicitor's office at Launceston. They were certainly there in the Civil War, for one of them raised a company at Morson for the King.

The slow march of generations is recorded on a double row of headstones in the churchyard of Bratton Clovelly, a mile from the farm. Mr Rice and I spent half a drowsy afternoon, scraping with the edge of a penny, to reveal yeomen Rices of Morson, steadily spanning the last century and a half. "There'll be an enormous quantity of Rices down there without stones to mark 'em," suggested Mr Rice, pointing at the blank turf.

It was Mr Rice's great-great-uncle James who attracted the Devil's interest, and the evidence is plain to see in the splendid old house, built like a medieval fortress. When Uncle James was well primed with liquor, he liked to stand at a bedroom window and shoot at the moon.

One night the Devil crept close behind him, but Uncle James was too quick for him. Whirling round, he was just getting the horned head into his sights as Old Nick fled downstairs. Uncle James came rumbling after him. As Nick darted across the kitchen, towards the front door, his hand touched the granite upright of the fireplace, and split it clean in two. Uncle James was so surprised that Nick got away.

The gaping crack in the pillar, 2ft wide and 1ft thick, is still there.

Everyone in these parts, of course, is convinced that the Devil died at nearby Northlew, and was buried at Okehampton. But I think he was hanging round again on the day, seven years ago, when the trouble started which is causing Mr Rice to leave.

One of his own cows knocked him down, and injured his leg. It did not seem much at the time, but he has never walked properly since. His lameness, increasing with years, has finally forced him to give in, at an age – 61 – when he could normally have hoped to farm Morson for another decade. "It's hard to think that I shall be the last," said Mr Rice, who left his native Devon as a teenage adventurer to seek his fortune in Australia. Fourteen years ago he came back to claim his heritage at Morson.

In a fashionable farming area Morson would be worth a fortune. But there will be no fancy £200-an-acre bids to the auctioneer tomorrow. This is a job for yeomen farmers, and I hope another race like the Rices gets it.

Ploughing Robsons chase that title
John Winter 29.8.63

Housewife Mrs Dorothy Robson, aged 25, and her husband, Raymond, 25, have important dates on October 30–31. They compete for the fourth time against each other in the National Ploughing match.

The Robsons, of Woodborough, Nottinghamshire, and 211 other local ploughing champions from all over the country will meet at Halso, near Taunton, Somerset. Two will be selected to represent Great Britain in the 12th World Plough championship on a state farm at Fuchsenbigl, near Vienna, in September next year.

Mrs Robson, a market gardener's daughter with years of ploughing experience, will compete on the first day in the wholework tractor class. Watching her, and her husband, who takes part in the semi-digger work class on the second day, will be Timothy, their two-year-old son.

The only other woman challenger is 37-year-old Miss Jean Burns, from Ramsey, Isle of Man.

There are only eight horse-drawn ploughs entered, three fewer than last year.

Farmer Tony hits jackpot

John Winter 11.10.63

Farmer Tony Emms, 45, who launched a five-year plan to earn £2,000 a year by working only six hours a day on his one-man farm, has hit the jackpot at the end of the third year. His family income, with his wife, Mary, acting as farm secretary, was £2,189 for the year ended last August. Now he has raised his target to £2,500 a year, and cut one hour off his working day.

He said yesterday: "People think I'm mad not to fill up my time by starting a secondary enterprise, and making more money. I think that's nonsense. I want leisure time to enjoy myself."

And so, on his 78 acres at Granborough, Buckinghamshire, he insists on remaining what he set out to be: an expert cowman, and nothing more. The farm is planned so that the 51 cattle can be fed, milked and managed by Mr Emms himself. All the field work is given out to contractors. "I work hard for three to three and a half hours in the morning and another two hours in the evening. The rest of the day is my own."

Until this year he and his wife and five children have taken a fortnight's holiday, by engaging a relief milker to run the farm. This year Mr Emms missed his holiday, but only in order to supervise the installation of a bulk milk tank in his dairy. He said: "It takes away all the graft of handling, cooling, and labelling 17 or 18 churns a day, carting them to the farm gate, and bringing back the empties. That saves me 70 minutes a day, and I reckon it's worth a holiday. Now I can get up after 6 am instead of 5.30."

Mr Emms, who is working to a five-year development plan, mapped out for him by economists of British Oil and Cake Mills, believes too many small farmers make the mistake of running too many lines of production. "The secret is to be a little man in a big way of business in one thing. My line is milk, and the limit of expansion is the number of cows I can handle myself – probably 55."

GENERAL

Where have all the milk churns gone?

John Winter 8.1.63

Thousands of milk churns have vanished from circulation in snowbound areas of the South and West.

The consequent shortage, coupled with a severe frost which brought traffic on by-roads to a standstill, made yesterday the most hazardous day for milk collection since the bad weather started. Two 1,750-gallon tankers, serving a bulk collection scheme at a group of farms in the Yeovil area, both ditched after skidding on icy surfaces.

Mr Toby Thwaites, Milk Marketing Board regional manager for Dorset, Somerset and Wiltshire said: "This is the first time since the snow came that we have had vehicles off the road. We have had to improvise churn collection from the tanker farms. One of the tankers has been pulled out, but the other will be stuck for a long time."

The vanishing milk churns have mostly been "hogged" by farmers stocking-up against further standstill. Yesterday the

National Farmers' Union appealed to them to play fair by their neighbours. An NFU official said: "Many dairies are collecting milk from several farms at a central point and replacing full churns with empty ones. Some farmers who visit the point soon afterwards are taking away as many churns as they can carry and leaving other farmers short, or entirely without.

Boys to the rescue   *John Winter 8.1.63*
Skiers from Hereford High School for Boys went to the rescue when hay, dropped from a helicopter to sheep belonging to Mr F B Powell, of Penlan, on the Black Mountains of Breconshire, fell short of its target.

Religious ritual and a 54-hour journey   *John Winter 1.4.63*
The suffering of sheep worried MPs yesterday.

First there were the 879 animals sent from Britain to Tunisia for ritual Moslem slaughter. Mr James Scott-Hopkins, Parliamentary Secretary, Ministry of Agriculture, said there had been only this one consignment in the last 12 months. He told MPs: "I see no reason why there should be any increase in this traffic." In March, Mr Christopher Soames, Minister of Agriculture, said that he had no intention of banning such exports.

Second came the plight of 457 sheep bought in Carlisle and taken by road to Dover. They took another 54½ hours to get to Paris and five were dead on arrival, said Sir Barnett Janner (Lab, Leicester North West). Sir Barnett asked for an Order to be made to prevent suffering to sheep exported from Britain. He added: "But for the fact that the sheep arrived in Paris in such shocking condition, they would have been sent on to Nîmes and Toulon."

Mr Scott-Hopkins said he would inquire fully into the matter.

The sheep were bought in Carlisle by a Dutch dealer in March. Sir Barnett said later that he had been told they were taken to Dover in double-decker lorries. They arrived too late to sail on March 9, a Saturday. There was no sailing on Sunday. They could not leave until Tuesday because of a dock strike in France.

Bogus dealers cashing in
John Winter 5.7.63
Bogus hay and straw merchants are reaping big profits by swindling farmers.

The crooks arrive in a truck, usually hired, and offer the farmer from £5 to £10 a ton for his crop. They pay cash for the first load. But when they return for more, they say they will pay later. Then they vanish. The tricksters can make anything from £200 to £300 for about £15 – the cost of the first load. Pickings for the experts can be as high as £1,000 a week.

They are cheeky, too. Their victims usually help to load the loot – and provide cups of tea.

One farmer, who was swindled, Mr H H Chaundey, of Chiselhampton, Oxfordshire, said: "I got £6 5s from one merchant when in fact he took £329 worth of stuff from me. The man left no real address and attempts by me to trace him have failed."

One Farmers' Union branch secretary, Mr John Davies, of Oxford, has opened a file on the bogus dealers. The names given by crooked "merchants" have been passed to him by farmers who were cheated. Now, when a dealer calls at a farm, a quick phone call to the union's office gives the farmer some idea if the man is a crook or not.

John Winter Reports on the final day of the Royal Show

John Winter 6.7.63

Further development and improvements to the Royal Agricultural Society's permanent ground at Stoneleigh Abbey, Warwickshire, were being discussed by society leaders as the first four-day Royal Show there ended last night.

On all counts, except weather – attendance, traffic handling, entries and trade stand business – the show exceeded the highest hopes of its promoters. When the gates closed last night a total of 111,916 had paid about £43,000 for admission in the four days. Nearly 50,000 cars, checked into five car parks, will bring in another substantial sum to set against the £250,000 spent on the show.

Of this, £100,000 was capital investment in ground services, which will not recur, and a portion of the rest will be saved next year by the absence of demolition costs and charges for moving and re-erecting the show on a new site. Even so, the show cannot hope to make a profit.

Mr Christopher Dadd, secretary, said: "We are bound to show a loss on paper, having regard to the exceptionally heavy expenditure. But the financial result is such that we shall be able to proceed with ground development on a substantial scale. If the gate had been poor the position would have been very different. As it is, we are most gratified by the response of the public who ignored the rain and mud and said it was a jolly good show, to which they would come again next year."

First attention in future planning will be given to more hard roads, and widening some already laid. This will help to reduce the mud problem, which has harassed show staff and visitors throughout four days of unsettled weather. Further treatment and consolidation of the newly sown grass will also contribute. Although record numbers of cars moved into the show area at an excellent speed, improvements to the traffic master plan, and modification to some car parks, will be recommended by Warwickshire police. No serious defects in ground layout have been revealed, but improvement will be considered in one or two sectors.

The show's last-day champion Mr Eric Osborne, 38, who won the Duke of Edinburgh's trophy for stockmen, was a London schoolboy evacuee who never went back. When he left a Buckinghamshire village school he got a job with cattle on the farm of Mr Corbett Roper at Lenborough. At 17 he was put in charge of the famous Roper herd of Dairy Shorthorns, which now numbers 350 head. It won two prizes at the show this week.

Rabbit disease

John Winter 6.11.63

Myxomatosis has reappeared among rabbits in East Bedfordshire and West Cambridgeshire.

Dartmoor has a beat-the-blizzard plan

John Winter 29.11.63

A plan to make sure that food can be got through to animals on Dartmoor in bad winter weather is ready to be switched into operation by the moorland farmers' organisation and the RSPCA.

The joint scheme is based on experience of last winter's big freeze, when animal protectionists took a hand in the land and airborne operations to save sheep, cattle and ponies. Farmers resented bitterly allegations of unpreparedness and neglect.

The plan, which has Ministry of Agriculture approval, is the sequel to a co-operative approach to the problem by Dartmoor Commoners Association and the RSPCA. It is based on a network of nearly 70 caches of hay at points all over the 200 square miles of moor. The sites have been selected for accessibility by tractor, and nearness to areas where sheep might be cut off by snowdrifts. From an operation HQ at Tavistock, a planning staff can take immediate steps to meet requests for help. The RSPCA will put an emergency staff of 12 men into the area within 24 hours.

1964

This was election year and Fred Peart became Minister of Agriculture – could the industry look forward to better times or just more controls? Christopher Soames, the outgoing minister, agreed a £30 million increase in guarantees at the annual price review. A milder winter led to an early completion of spring work followed by a good harvest, but the long dry summer still got blamed for watery potatoes that seemed reluctant to store. The Royal Show, now established at its permanent site, continued to expand. Farm accidents were down but there was talk of legislation about tractor cabs. We also learnt that the Aussie cricketers preferred milk to lager.

 ## POLITICS and FOOD

25 per cent – this is the rise farmers are demanding

John Winter 8.1.64

Farmers are demanding a 25 per cent increase – £100 million – in their incomes over the next three years. Mr Harold Woolley, NFU president, backed by his county branch leaders, announced this yesterday.

He said in London: "The time has come, at the 1964 price review, for recognition of the agricultural facts of life. Nothing less than a substantial uplift in farm incomes can enable agriculture to maintain the momentum of its efficiency and its ability to contribute fully to the national economy. It is our aim that the national net farm income should be increased by at least 25 per cent over the next three years and that a first and major step should be taken at the 1964 review."

Mr Woolley stressed that the increase would not mean a 25 per cent rise in either the Exchequer subsidy bill or food prices. The necessary improvement in farm incomes could be achieved if the Government reached its objectives of more stable food prices by co-ordinating imports and home production, plus better marketing arrangements.

Mr Woolley said that in the past five years farmers' costs increased 10 to 15 per cent, while prices dropped more than 4 per cent.

The Government's new policy proposals of standard quantities for home-produced foodstuffs were rejected by the Scottish NFU president-elect, Mr Michael Joughin, at the Oxford farming conference yesterday. The proposals, which limit to a stated quantity the amount of individual commodities on which the Government will offer price guarantees, is the quid pro quo Mr Christopher Soames, Minister of Agriculture, is asking farmers to accept in return for limits on competing imports. It had been expected that the farmers' leaders would accept this principle, with reservations.

Mr Joughin said: "I contend that the proposed standard quantities will not limit production but will only intensify the present economic squeeze on farming. They will aggravate the low income position and restrict the Government's ability to do anything about it at the price review."

Sir Alec tells of policy aims

John Winter 8.1.64

If Britain can avoid the wage-price spiral, the outlook is one of steadily rising average earnings and more purchasing power, of which a share goes to food, said the Prime Minister, Sir Alec Douglas-Home, yesterday.

He said at an NFU dinner at Kinross that there can be no doubt that industrial activity was quickening. He went on: "The farmers' market is the purchasing power of the industrial millions. It is to them that we in the countryside must look for the expansion of farming incomes."

He said the aims of policy in Britain are to provide a fair return to the efficient producer without a runaway bill for subsidies or penalising the consumer. He said there is a consistent agricultural policy which takes account of the interests of the farmer, the taxpayer, the consumer and the overseas supplier.

He went on: "No policy which will have a chance of enduring can neglect any of these interests. I hope we shall be able to reach agreement with our own farmers and with other suppliers of food on arrangements which will bring price stability to the market and provide fair shares in the growth in our consumption of foodstuffs."

Letter of the week

Mrs Joan Whittow, Orwell, Hertfordshire 25.1.64

Last week I took my small son to London and we had lunch at a well-known West End restaurant.

The choice was beef or lamb carved at the table. I asked if the beef was English or Scottish. This caused quite a flurry but eventually we found out, from the kitchens, that it was Argentine.

Now, I will not buy and eat Argentine meat. It annoys me to think they would not serve our own meat. Surely in such a restaurant our own beef should be advertised and served?

Union tackles bogies

John Winter 29.1.64

Two bogies which are sitting on every farm fence – limitation of production by price sanction and fear of markets being swamped by a few "concrete food factories" – caused the farmers' leaders considerable heart-searching yesterday.

The annual meeting of the National Farmers' Union in London agreed almost unanimously on qualified acceptance of the extension of the standard quantity system (which limits amounts eligible for Government guarantee), providing it allows adequate room for expansion and prices which will enable farmers to reach their 25 per cent income increase target in three years. This is a cornerstone of the new policy of Mr Christopher Soames, Minister of Agriculture, to co-ordinate home and imported food supplies, but it is now doubtful if he can reach agreement with overseas suppliers in time to introduce it this year.

But on the concrete farm threat there was an uneasy 3 to 2 division on asking the Government to warn off mass production intruders into the farming shires.

Mr Geoffrey Hart, Staffordshire, suggested that new style factory enterprises, using a few square yards of concrete for processing plants to convert feeding stuffs into animal products, would become an embarrassment, not only to the Government but ultimately the taxpayer.

He said: "At present we produce from our farms 76 per cent of the beef and veal eaten in this country. Now we hear of an enterprise being planned to produce 2,000 beef cattle a week, or one-thirtieth of our total output. In egg production, in which we are 98 per cent self-sufficient, one man is planning to run 6 million birds – enough to produce 6 per cent of the total."

Mr R B Thomas, Cornwall, called for the Government to limit the number of poultry or animals one producer could keep. The resolution asking for Government discouragement was defeated by 174 votes to 128.

Boost for farmers: £30m increase in guarantee accepted by union chiefs   *John Winter 16.3.64*

Farmers are expected to get an increase of about £30 million on the value of their guarantee in the price review settlement which will be announced on Wednesday. This is more than twice the previous highest increase since the end of food control.

The guarantees are the prices fixed in the review as giving the farmer a fair return. If the farmer's actual market price is lower the difference is made up by subsidy.

Leaders of the farmers' unions will recommend members to accept the award as an "agreed settlement" and this makes it an important political victory for the Government in its pre-election strategy.

If the unions had dissented, the settlement would have been imposed, as in three of the last four reviews. This would have released a tide of resentment among farmers, which would have resulted in many thousands of lost votes in the election and probably a number of marginal rural seats.

The Minister of Agriculture, Mr Christopher Soames, has had to preserve a delicate balance between doing justice to the farmers and exposing himself to charges of being over-generous to win their votes. At one stage during the four-week haggle, when he was pressing for a tougher settlement, the farmers threatened to walk out. At this point the Prime Minister is believed to have intervened.

After the unprecedented advance pressure campaign by the NFU, Mr Soames may consider himself lucky to have won its support with a £30 million increase. The union demand was for a 25 per cent income increase, or about £100 million over three years, with a substantial share of it this year. This was interpreted as a review target settlement which would produce at least an extra £40 million on income.

The increase in value of the guarantees is not an estimate of the increase in income farmers can expect in the coming year. But it is the only means of measuring the result of the review, and Mr Harold Woolley, NFU president, clearly believes it creates the opportunity for the income rise demanded.

He will have a tough job to "sell" his acceptance to the rank and file. With thousands going out of milk production every year because of the poor return, output is falling fast. The NFU demanded a "substantial" price increase and few milk producers will consider that the expected 2½d to 3d a gallon awarded in the review is enough.

The increase is unlikely to affect the subsidy bill as the Government will pass on the increase to the housewife.

Decline in dairy breeding herd halted
Peter Bullen 8.5.64

Mr Christopher Soames, Minister of Agriculture, found an election gift on his desk yesterday. In spite of many gloomy forecasts to the contrary, the March farm census figures show that the decline in the country's dairy breeding herd has been halted.

Short-term prospects for a bigger flow of home-grown beef are revealed, plus a heartening increase in the number of breeding pigs. But the milk supply position is going to be very tight for the next 12 months as there has been a 3 per cent drop of 81,000 cows. A summer drought could still cause an autumn shortage. The number of heifers and cows in calf are up by 55,000 – last year they dropped by 58,000 – and these will provide the reinforcements in the future. And these figures were collected before the price review which contained incentives for milk and beef production. Mr Soames can therefore count on the trends continuing.

The beef breeding herd was down 31,000, or 5.2 per cent, but there are quite a few more young beef animals in the "pipeline" which will be coming on the market in the next 12 months, aided eventually by the male calves from the dairy herd.

Breeding sows are up by 12,000; the ewe flock is up by 46,000.

Cereal growers may be in for a shock later this year. The barley acreage has jumped by 159,000 acres since last June to a record 4,307,000 acres.

This year Mr Soames brings in his standard quantity scheme to regulate home supplies. Given an average yield of about 28 cwt an acre, this huge area of barley in England and Wales will produce about six million tons of grain, which is only 500,000 tons short of the standard quantity set for the whole of the United Kingdom. If cereal growers go over this standard quantity they will incur disincentive payments for the first time.

Not entirely unexpected was the expansion in laying-hen figures. The number of fowls increased by 6½ million altogether, and this news comes during a week when output has already reached a record and producers are receiving their lowest-ever prices for eggs.

The number of men working on farms dropped by 4.5 per cent, or 14,500, but there were increases in part-time and seasonal workers.

Big help for the little man
Peter Bullen 28.5.64

Government backing for a plan to save hundreds of small farmers from being pushed off the land by big business factory farms is expected to be announced soon.

The scheme, worked out by the Agricultural Central Co-operative Association, is for a countrywide network of farm co-operatives. These will be made up of small units welded together to form an economic size and managed by the owners, who will make their own decisions on how they farm, but in a group instead of individually.

Such schemes are working very successfully on the Continent, and even in this country in the form of machinery syndicates. Since the beginning of this year machinery syndicates, formed by groups of farmers who buy machinery and equipment and share its use, have mushroomed. From 1957 to the beginning of this year about 480 syndicates had been formed. Since January 31 the number has jumped to 596. Their success is hailed by co-operative farming supporters as an example of how well British farmers can work together.

According to Sir Frederick Brundrett, chairman of ACCA, the choice facing most small farmers will be factory farming or family farming. Under the first, the farmer would probably be a small cog in a vast commercial enterprise, but by joining with a farm co-operative unit he would become a large cog in a small, but at least self-governing, enterprise.

As soon as the Government promises support for ACCA's plan a legal expert will be appointed to see how the new co-ops can be fitted into agriculture under present legislation. Then a small committee of experts, all well experienced in this field, will be named. They will work out the technical details of setting up the co-ops.

Farmers warn polling day victors
John Winter 7.10.64

Farmers yesterday said that they will expect whichever Party wins on October 15 to encourage a steady increase in home food production. They will also look to Britain's new Government to allow them to reach their three-year target of a 25 per cent income rise in the next two years.

These two steps are the essential basis of a 50,000-word "Green Book" of agricultural policy for the future* which is the National Farmers' Union's first comprehensive policy statement for 20 years. It touches every aspect of farming and food production, on which Government action can affect the well-being of farmers – and presents the role of British agriculture in the setting of a world food programme, which, Sir Harold Woolley, president of the NFU, says is going to be the paramount issue. Co-operation between major food trading countries is essential, the union says.

The case for expanding production from our own farms, already worth £1,800 million a year and filling half our food needs, is based on the help it can give to our balance of payments. It would also be a safeguard against shortages which could easily occur, such as the big drop in Argentine beef imports this year.

Among 77 recommendations, the statement deals with:

• *Beef.* Although a special study on the economic effects of factory farming will not be ready until next month, the union says intensive production, in which more animals are slaughtered at a younger age, could lead to a shortage. Last year 2.1 million cattle were slaughtered against 1.7 million four years ago. Farmers should be given the green light to expand production for five years.

• *Eggs.* The plans which the NFU and the Egg Marketing Board will present to the Government for curing overproduction may include differential payments for large and small producers.

• *Milk.* While milk should remain a guaranteed price commodity, the Milk Marketing Board's powers should be extended (it cannot at present fix retail prices), and a new grade of "plus milk", with more cream and solids, should be offered to housewives.

* *Marketing.* The NFU fears vertical integration moves by big food firms in which individual farmers would lose control of their own production. While keeping the door open for possible new marketing boards it urges farmers to band together in groups or co-operatives to market their own produce.

The union says that loss of farm land for building will increase unless steps are taken to control development. It urges measures to move population out of the South-East, a reversal of the Government's South-East Development Plan.

A shortage of farmworkers could imperil production unless the industry gets a policy of wages and conditions which will attract and retain enough workers.

Every General Election candidate will receive a copy of the statement.

* British Agriculture Looks Ahead, National Farmers' Union, 5s.

Soames tackles Minister

John Winter 3.12.64

Mr Fred Peart, Minister of Agriculture, and Mr Christopher Soames, his Conservative predecessor, had their first skirmish in Parliament yesterday. It came after Mr Soames had suggested that increases in wages, petrol tax and Bank Rate, and the import surcharge, would cost farmers well over £25 million a year.

Mr Peart, who had already estimated the cost of the recent wage award to farmworkers at £16.5 million a year and the Bank Rate rise at £5,250,000, refused

a request for a special price review to recoup farmers.

Mr Emlyn King (C, Dorset South): If we are to have two Budgets, is it not reasonable to have two price reviews; or, if not, an interim payment to farmers? Is it reasonable that farmers should be asked to lend money to the Government?

Mr Peart said Mr Soames was indulging in speculation. The procedure, as he knew very well, was for all the costs to be considered at the annual price review.

When Mr Soames asked if the Labour election promise to set up a new agricultural credit organisation meant that farmers would not have to pay the full 7 per cent Bank Rate, Mr Peart replied: "No."

The Agriculture Minister repeated his assurance that there would be no general increase in the retail price of milk this winter, and said he was considering urgently an application for an increase for the Channel Island grade to improve farmers' remuneration and encourage production.

Peart's gift to 10,000

Peter Bullen 23.12.64

Ten thousand hill farmers were given a Christmas present yesterday by Agriculture Minister, Mr Fred Peart.

He announced a change in the government grant they get for growing winter feed for their sheep and cattle. This will boost their incomes. Instead of present payments based on acreage the grant will be paid according to the number of sheep and cattle on the hill farm, in the form of supplements to the hill cow and sheep subsidies.

Mr Peart said in a written reply to a question in the Commons that he was making the change because he was concerned that assistance should be related more closely to need. For one reason or another many farmers getting the hill stock subsidies were unable to draw the winter keep grant. "The change I propose will overcome this difficulty and enable about an additional 10,000 hill farmers to share in the benefit," Mr Peart said.

Hill farmers in Wales and Northern Ireland particularly will benefit and the new scheme has the general consent of the National Farmers' Union, county agricultural executive committees and the hill farming advisory committee, said the Minister.

Actual amounts of the grant will be negotiated at the farm Price Review in February and will be paid during the 1965–66 winter.

CROPS and MACHINERY

Early spring means good start

John Winter 1.2.64

Spring in January has given farmers their best start for years with corn and grass crops. Many have begun to sow barley weeks ahead of normal, and the "early bite" on thousands of grazings promises to be the earliest ever. The only grouse: shortage of rain to wash in the fertilisers.

In the cornlands of East Anglia and the Southern Counties tractors are at work on cultivation and drilling of fertiliser and seed. The sowing of spring barley, usually done in late February or March, is in full swing on light land.

At Church Farm, Lavant, on the Sussex Downs, Mr Ernest Lock has already drilled 80 acres of Proctor barley, one-fifth of the total barley acreage he plans for his 1,330-acre farm. He said: "Weather and soil conditions have been so favourable we have been able to work without hold-up through December and January. After finishing winter wheat sowing we started with spring barley ten days ago and have drilled a lot more this week. We have sown a little in January before, but I can never remember a better start than this."

In Norfolk, too, barley is being sown in many areas of light land, and an NFU spokesman estimated that 2–3 per cent of the crop is already in the ground. Corn merchants report a big rush on spring seed, with a brisk demand for the newer barley varieties, like Cambrinus and Dea.

Pastures, on which farmers depend to give an early boost to milk production, are in excellent condition, though rain is needed in some areas. As far north as Lancashire cows have already been turned out. A herd of 50 on the Formby farm of Mr John Moores, Everton FC chairman, started this week to graze a field of rye. It was sown only last September, but is nine inches high.

Sweet news

John Winter 1.2.64

Sugar production from the 1963–64 home crop is likely to be 733,000 tons, an increase of more than 45,000 tons on the previous year. Average yield was almost 13 tons an acre.

The sugar content was 1 per cent higher than the previous year, and 45,000 tons more sugar was produced from 60,000 tons less beet. The best yields came from farmers in the Spalding, Lincolnshire, Ely, Cambridgeshire, and Ipswich, Suffolk, areas, all of which averaged more than 14 tons an acre.

A glut and low prices lead to carrot problem   *John Winter 21.4.64*

Farmers in Lincolnshire, one of the main carrot growing areas, were ploughing in whole fields of carrots yesterday because of a glut and low prices. At the same time they were angry over a BBC broadcast which advised housewives to buy foreign supplies.

The glut is caused mainly by the mild winter which has made other vegetables plentiful. In North Lincolnshire and Yorkshire alone more than 10,000 tons of carrots have been written off. Mr Ralph Kitson, secretary of the Isle of Axholme branch of the NFU, said that farmers were being offered as little as 30s to £2 a ton for their carrots. He added: "It would not pay growers to lift the crop and store it in the faint hope of selling later. In any case, the land is now wanted for other crops."

The broadcast which has angered the carrot growers was a radio programme for women shoppers. A speaker said foreign carrots were the best buy because there was less waste.

An official of Lincolnshire NFU branch said: "We checked in the shops and found that Lincolnshire-grown washed carrots compared very well with washed imported carrots. The Lincolnshire carrots cost the housewife only 3d or 4d a lb, but washed pre-packed foreign were 1s 3d a lb." He added: "It looks as if the housewife is quite prepared to pay up to 1s a lb to have her carrots nicely packed up in plastic."

Robot ploughs   *John Winter 22.4.64*

The development of robot ploughs which would work 22 hours a day "may not be far away," Mr J W Priest, president of the Institute of Agricultural Engineers, said in London last night.

New potatoes just in time

Peter Bullen 6.6.64

The first real flood of home-grown new potatoes this week has arrived just in time. Potato Marketing Board officials, who gambled on stocks of old potatoes holding out, sighed with relief when they saw the earlies on sale.

Last October the board forecast that "supplies and demand are roughly in balance," and after a December census said, "A serious shortage does not seem in prospect." Prolonged cold weather, damage to stored potatoes and/or a late new season could easily have brought a shortage and huge increases in prices to housewives. There would have been a gap if the board had not entered the market itself and bought thousands of tons under contract.

Now earlies are coming in from all parts of the country. Jersey growers have marketed about 20 per cent of their famous Royal variety; Pembrokeshire has cleared 7 per cent; Cornwall 3 per cent; and first liftings have been made in Cheshire, Kent and Lincolnshire. Rain is holding up lifting in some areas, particularly in Jersey. And the combination of warm and damp weather is increasing the threat of blight. Three outbreaks have been confirmed in Pembrokeshire, Cornwall and the Isle of Wight.

Housewives have to pay 10d to 1s a lb for the earlies in most parts of the country, but as supplies increase, day by day now, the price will come down. Old potatoes are 3½*d* to 5*d* a lb in the shops and producers' prices even went up by about £1 a ton in the last few days to £16–£20 a ton as demand increased when supplies of new potatoes were held up by the wet weather.

Biggest and best corn harvest

John Winter 4.7.64

Britain's biggest corn harvest is ripening in the fields. With good weather it will also be one of the earliest, and the starting price for farmers selling barley and wheat should be 10*s* a ton better than last year. Exceptional growing conditions following heavy sowings are expected to produce crops of over 7 million tons of barley and 3,750,000 tons of wheat.

Last year we got an estimated 6.6 million tons of barley and 3 million tons of wheat. But sowings have risen from 4.7 million acres to nearly 5 million acres of barley, and from 1.9 million to nearly 2,250,000 of wheat. Yields of both are expected to top the records set two years ago: 29 cwt of barley and 34 cwt of wheat to the acre.

Mr Charles (Micky) Norman, chairman of the working party on millable wheat and feeding barley, who received a knighthood in the Birthday Honours, yesterday estimated that the total corn harvest could well exceed 11 million tons. This includes oats, which are steadily declining in acreage, and mixed corn.

He also forecast, subject to continuing fine weather, that harvesting on the South Coast will start next week, a good three weeks earlier than last year.

His working party has set comparative prices for the opening of the new season of £19 5*s* a ton for feeding barley and £20 10*s* for millable wheat. These prices, for July and August delivery, are based on the prices of imported grain, and are an indicator to producers and buyers of home grain. The main reasons for the increases are the minimum prices for imports, under the new market stabilising scheme of Mr Christopher Soames, Minister of Agriculture, and the absence of any substantial carry-over from last year's crops.

Mr Norman said the big barley crop did not alarm him. An NFU official said that an early start to the harvest would help to spread marketing over a longer period.

Batting for Britain

Peter Bullen 16.7.64

Farmers may help England win the Ashes in 20 years time if they take up a suggestion in the Ministry of Agriculture's monthly publication, *Agriculture*, and grow willow trees to be made into cricket bats.

Berkshire farmer and forester Sir William Mount suggests that farmers might grow cricket bat willow and poplar trees in awkward corners and valleys on their land. He says: "Not only would they beautify the countryside but they would also provide the owner with a welcome cheque in the course of time."

The cricket bat willow can be worth from £12 to £25 when it has matured after 14 to 18 years, he says. But it needs a good site with free-running water. Poplars, not so demanding about conditions, take 20 to 30 years to mature. Then they are worth about £10 each.

Sir William, who milks 100 Ayrshire cows on his 850-acre farm near Aldermaston, also owns and manages 500 acres of woodland. He gives one warning: the cricket bat willow is extremely susceptible to frost in its first year. A severe frost late in May can cause havoc and it is unusual for the tree to recover. But when the tree is two or more years old a late frost appears to have little effect.

High yields from new wheat

Peter Bullen 12.9.64

New winter wheat, Rothwell Perdix, has averaged 42.07 cwt an acre compared with the Minister of Agriculture's estimate average yield for all winter wheat throughout the country of 31.7 cwt an acre.

Out comes the big stick

John Winter 6.10.64

More than £70 million a year is spent on farm machines and spares, but only about 4,000 farmers are prepared to pay three guineas a year to make sure that the machines they buy will do their job. This is the response to the Users' Testing Service, launched more than a year ago by the National Farmers' Union, in response to demands from members for a *Which*-type test of new machines.

The union, which organises the service in collaboration with the National Institute of Agricultural Engineering at Silsoe, Bedfordshire, estimated that 10,000 subscribers were needed to give a satisfactory service. Now it is preparing a "big stick" campaign to persuade more members to pay towards a testing service. The union fears that too many test reports already made are being circulated among farmers, and it has even had requests from branches for copies to be placed in branch libraries.

An NFU spokesman said: "If we do not get enough money it means that NIAE cannot test as many machines as they would like. The value of the service to farmers is shown by the tests already

carried out. They have resulted in at least one machine being withdrawn from sale, and many others modified. No fewer than 17 modifications were recommended on one machine."

Tests are made at Silsoe, or by NIAE experts on farms.

Drought puts work far behind

Peter Bullen 9.11.64

Stone-hard ground caused by near-drought conditions has put farmers weeks behind in their autumn cultivation work in the South. Apart from an area around Chichester in Sussex little ploughing or preparation of seedbeds has been possible.

But the months of sunny and dry weather have produced some unusual by-products. In many counties, but particularly Kent, Devon and Cornwall, fine second crops of strawberries – usual season May to August – have been gathered. Over a ton of strawberries has been picked in one area of Kent alone, and the fruit has been red, bright and sweet.

The countryside as a whole is looking smarter than ever before, as hedging, ditching and other maintenance work has leapt ahead. In the eastern counties and East Midlands ploughing and drilling of winter corn is only a matter of a few days' behind in most places. Conditions have been difficult due to the dry, hard soil, but not impossible.

Farmers in the North are well on schedule because of the slightly higher rainfall. A lot more wheat than usual has been sown in the North, said an NFU observer yesterday.

Machinery exports booming

Peter Bullen 18.11.64

Britain's farm machinery exports this

year are running at an even higher rate than last year's record levels. In the first eight months they were £4 million, 3 per cent up on 1963.

This was announced yesterday by the Agricultural Engineers' Association and it coincided with an announcement by Ford of a new range of tractors, which should capture even more export orders. Mr Geoffrey Buckley, general manager of Ford UK Tractors, said: "Our short-term forecasts indicate that world-wide tractor sales in 1965 will set records. By 1970 we expect the world tractor market will increase at least 18 per cent from 10½ million to nearly 12½ million tractors."

Last year Ford exports totalled £157 million, 25.4 per cent of which came from tractors, and they had completely reorganised their tractor production to improve on this, said Mr Buckley. The old tractor factory at Dagenham was being modernised for car production and two new factories had been built; a £20 million plant at Basildon, Essex, and a factory at Antwerp.

The range of four new tractors replacing the present three sizes has taken more than three years and £35 million to develop. The new tractors are still on the secret list but details will be announced a week today.

Mr Buckley said Britain gained 63 per cent of world tractor exports last year and £140 million of the record £165 million-worth of Britain's farm machinery exports were tractors and tractor parts.

Air Vice-Marshal F L Hopps, chief executive of the Agricultural Engineers' Association said yesterday that the Government's proposed rebate on indirect tax for exporters would probably favour agricultural engineering more than most industries. This was because of the high proportion of its output which is exported – 64 per cent last year.

Fords' new range

John Winter 25.11.64

A new range of tractors is launched today by Fords. An outstanding feature is the automatic transmission which is an optional extra on the three larger models (£125 or £150, according to model).

Blame the sun for these soggy spuds

John Winter 22.12.64

There will be two crisis points for housewives all over the country as they prepare the dinner of the year on Friday. One will be the moment when the turkey is done to a turn; the other when the potatoes begin to disintegrate in the pan and float away into a soggy, watery mash.

This stubborn refusal to stay solid when boiled is a chronic weakness of potatoes this year, and it's no use blaming the farmers, greengrocers, fertilisers, or even the Potato Marketing Board. It's the weather. The glorious summer and warm, dry autumn played a shabby trick on potato lovers. They produced a crop which, by and large, is excellent in appearance – clean, free from skin blemishes or cracks, and of even size. The trouble lies under the skin.

Owing to lack of rain in the later stages of growth the average potato this year contains at least 5 per cent less moisture than usual – and that is causing a lot of frustration in the kitchen.

Potato sleuths of the Marketing Board and scientists at two Cambridge research stations are investigating the problem. Tests show that disintegration in boiling is common this year to crops of all varieties grown in all parts of the country. It does not occur in early potatoes, which are eaten at an immature stage of growth, but it is a hazard to all maincrop varieties, in which growth is completed before they are lifted.

The critical development period is in the late summer and early autumn, when the tubers are bulking up. In a year of average rainfall potatoes consist of about 80 per cent water and 20 per cent dry matter. Some varieties have more water while others, like King Edward, have higher dry matter which gives them, when boiled, the floury texture that most people prefer. This year the water content is down to between 70 and 75 per cent and it is the extra dry matter which causes the maddening behaviour of potatoes in the pan.

The dry matter consists largely of starch, which turns floury when boiled. This year there is so much of it that it breaks down the cells of the potato and bursts out into the water.

"I suppose if we knew what kind of weather we were going to have we could advise farmers which varieties to grow," said a scientist at the Institute of Agricultural Botany. "If a farmer knew a long drought was on the way he could arrange to irrigate his crop. But applying the equivalent of one inch of rain a week, which is what would be needed from mid-July to the end of August, would make potatoes terribly expensive."

For the housewife who hopes to serve potatoes which are dry, firm and floury the boffins have this advice: steaming, or very gentle boiling, is better. The important point is to watch them constantly as cooking nears completion. The disintegration starts suddenly and potatoes break up in a matter of seconds. So whip them off at the first sign of "floating" and drain them. If they have been boiled slowly they should be cooked right through.

Or why not save all this anxiety by roasting them or baking them in their jackets?

 # LIVESTOCK and POULTRY

Boost for Channel Isles milk

Peter Bullen 15.2.64

The first step to merge the two Channel Islands cattle breed societies will be made next week. This would give a boost to the production of the rich, creamy milk which is now in short supply.

At the annual meeting in London on Wednesday of Quality Milk Producers, the organisation which promotes the sale of milk from the Guernsey and Jersey breeds, the move will be proposed by Mr

F Raymond Stovold, of Godalming, Surrey.

Mr Stovold, council member of the English Guernsey Cattle Society and a former president said yesterday: "By merging our administrative, sales promotion and publicity efforts we would save thousands of pounds a year, and become a much stronger body as well. At present the work is being done by three organisations in three offices when one headquarters would do, Agriculture House, London, for instance."

Many Channel Islands producers are backing Mr Stovold and any practical suggestions for more co-operation between the two breeds will get a lot of support at Wednesday's meeting. A leading member of the South-Eastern Jersey Breeders' Club, Mr J N Birch, of Horsham, Sussex, said he hoped to be at the meeting to support the merger plan. He added: "In my view it is wrong for any agricultural organisation to ask for more Government help until it has effected its own economies. One CI milk organisation would be much more economical."

At present the country's 8,500 CI producers are unable to meet the demand from housewives for the milk despite an annual output of about 110 million gallons.

Robot pig-feeder   *John Winter 10.3.64*

A robot on a farm at Little Bookham, Surrey, is bringing Large White cross Landrace pigs to bacon weight in 140 to 147 days – more than a month before hand-fed pigs from the same farm go to market.

Two of them took only 119 days before being ready for market and 95 per cent are being graded AA and AA+, the two top grades. The hand-fed pigs average 180 days to bacon weight and food costs are about 15*s* more per pig with only 80 per cent grading AA and AA+.

For the past two years Mr F Seabrook has been carrying out trials of the two methods. All pigs are bought as eight-week-old weaners, weighing some 40lb each. Each year 900 are fed by the robot and 300 hand fed twice a day.

Secret of the robot's success is probably the fact that it gives the pigs a 24-hour service. It is a machine which can be preset to operate at regular intervals day and night. Up to 12 weeks old the pigs receive a feed every four hours and thereafter every eight hours.

Mr Seabrook says that the machine-fed pigs receive their food in smaller amounts, but more frequently, and utilise it more efficiently. In addition their digestive systems were smaller than pigs which had big meals and this had a bearing on the amount of usable meat in the carcass.

Beef drain to Europe worries butchers   *John Winter 7.4.64*

The present beef shortage is likely to get worse and prices will rise even higher in the next month or two. This was yesterday's forecast from 600 butchers, in conference at Scarborough.

They claimed that export of home-bred beef and dairy cows for slaughter on the Continent is aggravating the position. The conference of the National Federation of Meat Traders Associations passed resolutions deploring the trade to Europe and calling for the subsidy on the animals to be repaid to the Government. The butchers said that because so many of the cows were in-calf, beef production as far ahead as 1966 would be affected.

West Suffolk butcher Mr A Lacey said that foreign buyers were raising the price for British cattle with the result that our housewives were going to pay more for meat.

Mr W Goodfellow, of Halifax, who proposed the resolution calling for the subsidy on livestock to be repaid by the exporter, said: "It is not sound policy for the taxpayers to pay subsidy on livestock which are to supplement another nation's larder."

Mr Tom Tasker, the federation president, said that the butchers had tried to warn Mr Christopher Soames, the Minister of Agriculture, that our meat supplies were very finely balanced.

Beef by the ton: new farm factory will hold 10,000 calves

John Winter 20.4.64

The first 230 calves in a revolutionary "beef for the million" project are munching their way towards the butcher's slab on a remote 60-acre farm in Suffolk. They are the first to go into a barley beef "factory", built to hold 10,000 head, with a weekly output of 200 beasts, or more than 50 tons of beef.

The £750,000 enterprise, started at Brown-street Farm, Old Newton, near Stowmarket, by British Beef, a subsidiary of Union International, the meat combine, is reminiscent of an American feedlot.

The animals are housed in strawed pens in 340yd-long sheds. Ten are being built, which, end to end, would stretch for more than two miles. They are on both sides of a central gangway. Construction is simple and cheap. And the rearing methods should rebut any charges of cruelty from opponents of factory farming.

The cattle stand, or lie, comfortably with ample space on deep straw laid on the earth. They have constant access to a food trough at the front and a water tank at the rear. Each pen has a rack of unlimited straw, which they munch to get roughage. There is no concrete and no slats. The entire construction is of galvanised metal sheets, interspersed with transparent plastic, fixed to a framework of rough-hewn poles.

A wide gap along the shed ridge, and others along both sides, admit all the light, air and even rain that any animal lover could desire. Yet the pens are so built that the cattle can always get shelter, even in a gale.

The whole "factory", including a feed mill and grading shed, will cost only £10 a head of animals housed, a fraction of the cost of many cattle units in this country. The farm also breaks tradition by bringing in hundreds of suppliers on a contract-farming basis. Calves are being drawn from dairy farms in many English counties, and as far north as South Scotland. Producers supply them on annual contract at £8 to £10 a head.

Animals and humans: an astonishing affinity

John Winter 29.4.64

The relationship between men and animals is going to be a central problem of the factory farming revolution which is sweeping through agriculture.

This has nothing to do with the charges of cruelty, or overcrowding, which have been made against some of the new intensive production methods. It is a question of the social needs of animals, especially cattle, which are the latest subject of factory farming development in Britain.

Two incidents at one beef farm, run on the feedlot principle, show how little we know about cattle psychology.

Here, 400 cattle are housed in a covered yard, built at considerable expense, and providing them with every comfort. The first incident was caused by the high-pitched stutter of a motor-scooter being started up near the animals; the second, at night, by the headlights of a car. As it came along an adjoining road it sent a beam of light flashing into the building, through a gap between sliding doors.

In both cases the result was the same – panic, and a stampede.

The sudden, relentless movement of 150–200 tons of beef on the hoof is a powerful and frightening force. In both cases the animals burst open the wall of the yard, as though it were made of matchwood. When you recall that beef factories holding up to 10,000 animals arc now being built, you appreciate the size of the problem. Ten thousand steers and heifers is about 4,000 tons of beef on the hoof.

The significance of the scooter and headlight incidents is that both occurred when no stockmen were with the cattle. If only one had been present, the result could have been different. This hunch is based on the studies of Stephen Williams, general manager of the English and Scottish farms of Boots, the chemists, and one of our leading authorities on farm animal psychology. From his observations on the firm's farms at Thurgarton, Nottinghamshire, he believes that cattle have a definite need of regular contact with their stockmen, although the part this plays in their behaviour is not fully understood.

Mr Williams has recorded these strange happenings among the Thurgarton herds.

1. More cases of animals straying, or breaking out of fields, occur on Saturday afternoons and Sundays, when the staff are not around, than in the whole remainder of the week.
2. When Mr Williams takes parties of visitors into fields where cattle are grazing, the animals which stand closest to the strangers are always the same ones.

Both these phenomena, he says, are closely linked with the social order of precedence, which is established in any large group of animals, and which is well known to farmers. In this "bunt order" there are the top dogs, or bosses, which exert a bullying domination over all those lower down the scale, which descends progressively through the middlemen, to the underlings.

In the weekend straying, it is not, as you might expect, the bullies which break out, but the underlings. Probable explanation is that, in the absence of stockmen, the overlords grow more aggressive than usual. Life for the tail-enders becomes so unbearable that, in their efforts to get away, they stray from their regular haunts.

When a party of visitors enters the field, the mere presence of humans, especially any who are known to the animals, or have loud voices, immediately reduces the supremacy of the overlords, and correspondingly raises the status of the underlings. The bosses make for remote parts of the field. The underlings still dare not challenge them by being too close, and so they find themselves in the most obvious clear area, right in front of the party. This positioning may be almost involuntary, or it may occur because they feel happier in close proximity to humans.

"It doesn't matter how often a visiting party moves its position, thereby causing a regrouping of the cattle, the same animals are always closest to us," said Mr Williams. "If I carried a camera they would always be in the middle foreground of the picture."

While the influence of the bunt order may not be fully understood, it is clear that the regular presence of humans affects it profoundly. How differently, then, will animals behave when they see, and hear, much less of their stockmen? (It is not clear whether sight, or sound, has the greater effect.)

"Factory farms will clearly need stockmanship of a very high order, if they are going to get the best results from their animals, and avoid accidents," said Mr Williams. It is possible that loudspeakers, playing music, or recordings of a familiar voice, may play a big part. Closed circuit television, showing a picture of the cattle pens to the stockman sitting in his home, or office, may also be valuable."

Factory farmers may have to re-learn gimmicks which were known to the Romans. They chose their cattlemen for the strength of their voices, and pigmen, or porculators, for the sweetness of theirs. It was found that if they sang at feeding time, the pigs did not fight at the trough and presumably derived greater benefit from their food.

Letter of the week – "I'm laughing"
W Tennant-Eyles,
Swanboune, Buckinghamshire 1.6.64

Who is sorry for the butchers, except the butchers?

Any farmer who has had fatstock sold at barely above cost price, owing to butchers forming buying rings at markets, is getting a little of his own back now. I enjoy seeing foreign buyers outbidding local butchers. It is now the farmer's turn to laugh, and I am certainly laughing.

Farming is the most maligned industry in Britain, from which butchers have made a fat living for years.

New exports drive launched
John Winter 20.8.64

A £20,000 cargo of pedigree livestock sails from Felixstowe today to open a new market for British agriculture – behind the Iron Curtain.

The cargo comprises 40 head of Ayrshire, British Friesian and Hereford cattle, 30 Landrace and Large Black pigs, and 25 Kent and Romney Marsh sheep. With more than 200 head of chickens and turkeys and exhibits of equipment from ten firms they will present a comprehensive picture of British agriculture to Czechoslovak farmers at the Prezov Show, the country's top farm exhibition, next month. The show dates have been put back three months to accommodate this first all-British exhibit.

Mr Rudi Sternberg, in his position as chairman of the Livestock Export Council, which has organised the operation, said: "In the next few years I believe agricultural exports, worth many millions, will be the biggest item in our trade with Czechoslovakia."

Best bull in the world dies – and leaves 4,000 offspring
Peter Bullen 3.9.64

Histon Dairy Premier, grand old man of the oldest cattle breed in Britain, the Dairy Shorthorn, was being mourned yesterday. The breed's most famous bull had just been put down at Cambridge artificial insemination centre. His 18-year record has been impressive. His progeny – well over 4,000 – can be found all over Britain and in Australia, New Zealand and America.

Yesterday the Shorthorn Society issued a two-page obituary notice in his honour, which said: "Has there ever been a bull like him? . . . there can be little doubt that he has sired more Dairy Shorthorns than any other bull. In fact, it could well be that his progeny outnumber those of any bull of any breed. . . . His influence in the breed will be felt for very many generations to come."

At the Cambridge AI centre they were describing him as "the best bull in the world . . . docile and obedient." At least three of his daughters have produced more than 100,000lb of milk each. He was bred by Mr J Stanley Chivers, of Histon, Cambridge, a past president of the Shorthorn Society.

French invaders get a vote of confidence
Peter Bullen 9.9.64

One of the most eagerly awaited reports from the Ministry of Agriculture is to be published on Monday – the first comprehensive survey of the progress of the Charollais bulls imported from France in November 1961.

The French "invasion" was organised by the Ministry to see if these heavy, quick-growing bulls could help to produce beef more efficiently than some home breeds when mated with our dairy cows. Progeny of these crosses have now been reared under intensive methods and have gone to market. The report of progress so far is being published in the Ministry's 1s monthly publication, *Agriculture*, on Monday.

In the main the report gives the Charollais a vote of confidence. Both steers and heifers from cows mated with the French bulls put on about 5 per cent more weight every day than their British counterparts. And they did so despite eating much less. The report says: "Consumption of starch equivalent per lb of liveweight gain was almost 10 per cent less for the Charollais crosses."

Pure British Friesian steers were not outdone on liveweight gain, but did not quite match the Charollais-Friesian steers on carcass grading. The French influence also had a notable effect on increasing the percentage of meat remaining after the animals had been slaughtered and the waste products removed.

When results of the Charollais crosses still being reared by traditional extensive methods are completed next spring, the Ministry will decide whether more Charollais can come over to establish the breed here.

 # EMPLOYMENT

Letter of the week – the Beatles
B W Reynolds, Waddon, Surrey 10.2.64

You report the secretary of Surrey National Farmers' Union wanting consumers to pay an extra ½d a pint for milk.

What does he think we are? Let him advise his members to get up a little earlier; do a little work themselves; sell the Jags and buy Minis. We do not want to pay any more. Let farmers be satisfied with a fair return and not imagine they can equal the Beatles.

Reply to letter of the week
Mrs Eileen E Prout,
Elmore, Gloucestershire 13.2.64

May I write a word in favour of the farmer? My husband gets up at 6 am and finishes work outside at 8 pm, on seven days a week, 52 weeks a year – and has not had a holiday for 11 years.

None of our farming neighbours owns a Jag, and many take their wives to church in the same conveyance as they take their calves to market. We do not imagine we can equal the Beatles – they only serve the public for a few months. Wake up, Mr Reynolds, this isn't 1864.

Mr B W Reynolds' letter on Monday prompted a record response from readers all defending dairy farmers as above, and pointing out the decline in their incomes which has led to 40,000 quitting milk production in the past ten years.

Chance for young men with £750

John Winter 15.4.64

Young men with limited capital who want to find a way into farming were offered a key by the chairman of the Land Settlement Association, Mr Jocelyn Gibb, yesterday. A number of the LSA smallholdings, average size five acres, change hands every year and the association was always on the look-out for new applicants of the right calibre, he said.

The qualifications are:

1. Minimum of five years' experience, which can include up to two years at a farm institute or agricultural college.
2. Savings of at least £750.

Mr Gibb said: "The working capital needed by a newcomer is now well over £2,000. Up to three-quarters of this can now be borrowed but we like, as far as possible, our new tenants to have at least £750."

Men with the right qualifications who do not mind which part of the country they begin in could be given a holding within a few months. For holdings in particular areas there was a two- or three-year waiting list.

Last year the LSA's 710 tenants chalked up an average income of £885. Of these, 262 made more than £1,000, 29 made £2,200 or more and a few even passed £3,000 – all from a few acres. At Sidlesham, Sussex, 114 tenants set up an all-time record for an estate by making an average of £1,400 per unit. Their secret was "jolly good, hard work," said Mr Gibb.

The LSA, which markets all its tenants' produce, sold £2,251,585 worth for its 710 tenants last year. Pigs made £690,000, poultry £56,000, eggs £311,000, produce £1,136,000, miscellaneous £56,000.

More pay

Peter Bullen 31.7.64

A pay claim by farm workers will be replied to by employers on September 15. The National Union of Agricultural Workers yesterday asked for a £2 10s-a-week rise, making workers' minimum wage £12.

Letter of the week – bring pay into line

A Horter, Slough, Buckinghamshire 3.8.64

Agricultural and horticultural workers are still expected to exist – I won't call it live – on a basic wage of under £10 for a 45-hour week. Surely it is time someone gave us a break and brought us into line with industry. Most of us are just as highly skilled in our own right as any industrial worker and work in far less congenial surroundings.

One final thing. Any pay award made to most other industries is back-dated by anything up to 1½ years. Ours is always dated forward by a month or three. Why?

Pigmen go back to school

Peter Bullen 1.9.64

An experimental course to train pigmen for the growing number of large, intensive units starts next week. The ten-day residential course at Kesteven Farm Institute, Lincolnshire, will include papers on pig keeping, demonstrations and informal discussions. Visits to large pig farms will also be arranged to study different management techniques, and a bacon factory will be inspected.

Ten pigmen are paying £10 each for the course which, if successful, could lead to larger regular ones. Mr G R Brent, the institute's lecturer in pig husbandry, is in charge and visiting speakers include Mr Ronald Major, national pig adviser for the feed firm which is organising the course with Lincolnshire's County Education Committee.

One of the organisers said last night: "We hope the farmer who employs the type of pigman wanting to take the course will give him time off and perhaps even pay him. There is a great need for this type of training."

Claims rejected

John Winter 16.9.64

Claims for an extra £2 10s a week for Britain's 535,000 farm workers were rejected yesterday by employers at the meeting of the Agricultural Wages Board. They said the increase, if granted, would cost the industry more than £50 million a year.

On push-button farm where workers make £1,000 a year

John Winter 26.9.64

The cattle and pigs on "Pushbutton Farm" are tended by £1,000-a-year farmworkers who run their own cars. Electrical automation has taken the drudgery out of their work. But the pushbutton farmer, Earl De La Warr, says that landowners with "a little more capital than his neighbours" should not aim at running a show-place farm. They should develop one that was a little ahead of neighbouring farms so that it could be of use to them.

The electricity bill for the 776-acre Buckhurst Farm on the earl's estate at Hartfield, Sussex, is £554 a year. The farm has all-electric mixing, delivery and feeding of the rations for 140 Jersey cows and 1,000 heavy hogs. In 17 years the total manpower has been cut from 35 men to ten, including the manager, Danish-born Mr Arnold Christensen, and a 93-year-old pensioner. Mr Christensen said: "Our electricity bill may seem high, but it covers 38 electric motors and all lighting and heating, and it is less than half the cost of a stockman."

With a basic wage of £15 a week, a free house worth £3 a week, holidays, days off and other incidentals, the cost of one man at Buckhurst is £1,246 a year.

"We aim to pay our men such a wage that they can have the things people are entitled to in a civilised society," said Mr Christensen. Good working conditions attract a high-class labour force, most of whom move on after a few years to become farm managers, or farmers in their own right.

"Feeding and cleaning out pigs by traditional methods is not a very pleasant job," said Mr Christensen. In any case it costs seven or eight times as much to wheel a barrow a given distance as it did before the war. Yet electricity has increased by only 20 per cent in 20 years.

The farm is being used as a working model of electrical automation by the

Electrical Development Association. Next week it will also be studied by the Government committee which is investigating the welfare of animals on "factory farms".

Rachman slur angers farmers

John Winter 14.11.64

Farmers are furious with the Government for classing them with racketeering urban landlords who evict tenants at the drop of a hat.

More than 140,000 farm workers live in tied homes owned by their employers, and the National Farmers' Union is smarting under the "Rachman" slur cast by the Protection From Eviction Bill. It is rushing out a statement of the farmer-landlord's position to MPs before the Bill is debated next week. It hopes to get an amendment deleting farmworkers' tied homes from the Bill.

An NFU spokesman said: "What we particularly resent is that farm cottages have been singled out, while the Bill makes no mention of the vastly greater number of other service houses." He added that the number of removals attended by any suggestion of hardship is insignificant.

Death toll cut again

Peter Bullen 30.12.64

The 1964 death toll on Britain's farms was below the 100 mark for the second year running, showing that the Ministry of Agriculture's "Keep Farming Safe" campaign has had a heartening impact since it was launched nearly two years ago. Deaths totalled 141 in 1961, 127 in 1962, 97 last year and 99 so far this year.

Mr George Wilson, the Ministry's chief safety inspector, said yesterday that he felt encouraged that last year's drop was not a flash in the pan. "The reason probably is that the safety regulations have had their effect. The only way to reduce the present toll substantially will be by greater care by people who live or work on the land."

The worst aspect of this year's figures is again the large percentage of deaths caused by tractors overturning. Thirty died in this way, and nearly every case was due to the human factor rather than any unforeseen mechanical cause.

Mr Wilson said: "Another matter of concern is the large number of children killed. For the past three years 16 have died each year, and 17 in 1964. Old people, too, are a problem with 12 people over 65 being killed on farms this year. Obviously, more care must be taken of children who do not appreciate dangers that exist on farms. The trouble with old people is that they think they can do things they did when young, like riding on top of loads."

Careless handling of guns caused several deaths. Six people were electrocuted, the main causes being do-it-yourself wiring or bad maintenance of installations.

Many farmers are not bothering to get combine harvesters, binders, pea and bean harvesters, pick-up balers and windrowers converted in accordance with machinery regulations which come into force on Friday. Mr Wilson said: "We know farmers are not bothering about this yet because of the large number of safety guard conversion kits which agricultural engineers and dealers still have on their hands."

Safety cabs

Peter Bullen 31.12.64

Britain's 350,000 farmworkers want new Government legislation to make the fitting of safety cabs on tractors compulsory. The National Union of Agricultural Workers is appalled by the large number of deaths each year caused by drivers being crushed by overturning tractors. As revealed in Farm Mail yesterday, 30 out of the 99 deaths on farms this year were caused by overturning tractors, and similar accidents cause about a third of farm fatalities each year.

At the moment the National Institute of Agricultural Engineering at Silsoe, Bedfordshire, is coming to the end of a series of tests on tractor safety cabs in an attempt to lay down minimum safety requirements. An NUAW spokesman said yesterday: "When a minimum standard for testing these cabs has been worked out, we will certainly try to get the Government to make a statutory requirement that the cabs are fitted to tractors – as the Swedish Government has done."

Mr George Hook, NUAW safety officer, said, "We think a lot could also be done in training and supervision of tractor drivers, particularly younger men, to prevent these accidents."

Farmers were also concerned, said a National Farmers' Union spokesman. The union was co-operating with the British Standards Institute, the Royal Society for the Prevention of Accidents and the NIAE in an attempt to keep accidents down.

"We would like to bring about a better standard of safety without adding to the vast number of compulsory regulations if possible. We would certainly encourage members to fit safety cabs on tractors – when a really safe cab is decided upon. We feel this is probably a more logical way of tackling the problem," he said.

The Ministry of Agriculture, which has issued several warnings about the dangers of tractors overturning, has published a special safety leaflet which warns tractor drivers not to drive too fast, and to take special care near ditches, banks, on steep hillsides, and on silage and manure heaps.

PEOPLE in the NEWS

Pigeon pie is sport and business
Peter Bullen 5.2.64

Farming's "have gun, will travel" man, 34-year-old Mr Eddie Williams, will be claiming more victims in East Anglia tomorrow. With thousands of farmers, gamekeepers and other expert shots he will be taking part in mass shoots organised by the Ministry of Agriculture in the Eastern counties, North, Midlands and Wales. The shoots are part of the Ministry's war on the five million or so wood pigeons which account for millions of pounds' worth of crops each year.

Mr Williams, of Purley, Surrey, is the wood pigeons' No 1 enemy. Last year he travelled 35,000 miles, shooting in seven counties, and bagged 14,000 pigeons, probably the highest individual tally in the country.

He said yesterday: "I do it as a hobby in my spare time. But there is hardly a week in the year when I don't have two days shooting. I pay farmers 1s each for the birds they shoot and sell them, with the birds I have shot, to an agent who exports them to the Continent. Most farmers can get only 3*d* or 4*d* each for these birds normally."

The Ministry's mass shoot will continue for the next six weekends and will be followed by a three-month nest destruction campaign in the summer.

80 today, but Bill will plough on

Peter Bullen, undated

Mr Bill Hudson, ploughman extraordinary, celebrates his 80th birthday today. And waiting for him is a special "present" from his boss – a new £850 tractor. He will use it when he ploughs the 250 acres of arable land at West Stoke Farm, near Chichester, Sussex.

His boss, Mr Robert Mason, said last night: "I am giving Bill a small personal gift, but this tractor is to mark a special occasion. He must be the only man in the country who has been driving tractors for the past 50 years, and is still working full-time. He is a wonderful worker and has worked for my family for over 35 years. This is to show how much we appreciate his dedication to the job, and to agriculture."

Mr Hudson works more than 40 hours a week for most of the year, and longer at harvest time, when he carries on until nine or ten in the evening. Every year he does all the ploughing on the farm. He started to learn the art when he left school before the turn of the century. He used a team of horses under his father's tuition, and his first week's pay was 4*s* 6*d*.

Mr Hudson was thrilled with a pre-birthday trial on his new machine. "These new 'uns are much more comfortable and a lot easier to start," he said. "The seat is so comfortable on my new one I don't want to get out of it."

Retirement: "Hard work keeps me going," he said.

Wallace's whistle-stop tour

John Winter 18.6.64

Mr Wallace Day, the West Country farming firebrand, yesterday passed the 3,000 mile post in his whistle-stop tour to win votes for his election to the Milk Marketing Board. In three weeks he has canvassed farmers in counties from Cornwall to Yorkshire. And, as 105,000 milk producers started voting for a special member for England and Wales, Mr Day was confident the poll would be far above the low percentages of the past.

He told me: "I have concentrated on three simple issues. One: it is time there was a change in board personnel. Two: we need a milk production regulator who will answer the Minister of Agriculture's arguments that a price increase leads to an embarrassing upsurge in production. Three: the small milk producer deserves stronger representation on the board."

Energetic campaigns have also been conducted by the other two candidates – Mr Stanley Morrey, of Wiltshire, and Mr George Pigott, of Hertfordshire. Mr Morrey, who is nominated by the board, claims the support of a majority of NFU county branches.

All electors have received their voting papers. The poll closes on July 3, and the result will be declared a few days later.

Farm Mail in Parliament:
Lady Barnett in farm inquiry

John Winter 30.6.64

Lady Barnett, the TV personality, is to be a member of the Government committee inquiring into factory farming. Her appointment was announced in the

Commons yesterday by Mr Christopher Soames, Minister of Agriculture.

Lady Barnett, a magistrate and a doctor of medicine, lives at the Leicestershire village of Cossington, in a rich farming area where intensive methods are used.

Chairman of the committee, a technical one, is Professor F W Rogers Brambell, head of the zoology department of the University College of North Wales, and the other seven members are all leading figures in agriculture or animal studies at universities. They have to examine the conditions in which animals are raised and kept, advise whether standards should be set in the interests of their welfare, and, if so, what they should be.

Lady Barnett's appointment brings two women into the factory farming row, which led to the inquiry. It was precipitated by London housewife Mrs Ruth Harrison in her book *Animal Machines*, which condemned modern intensive methods of raising livestock.

It's a home win for Britain

Peter Bullen 9.7.64

Five hours of tension ended in clear-cut victory for British young farmers' teams competing for the *Daily Mail* Gold Cup for cattle judging at the Royal Show at Stoneleigh, Warwickshire, yesterday.

Winners with 1,273 points out of 1,600 were the Welsh team of four dairy farmers' sons – the first Welsh victory in the international dairy cattle judging competition's 35 years. England was only just behind with 1,261 points, Scotland scored 1,228, US (last year's winners) 1,177, Jamaica 1,166, Ulster 1,094, Australia 1,089 and New Zealand 990.

News of the Welsh victory came just before the Duke of Gloucester arrived at the young farmers' stand to present the *Daily Mail* cup. The eight teams had inspected cows of the Ayrshire, British Friesian, Dairy Shorthorn and Guernsey breeds. Each team member then had to place the animals in order and explain to the judges in two minutes the reasons for his or her choice.

The winning Welshmen, who said: "We had hoped, of course, but never thought we would pull it off," are Gareth Thomas, 22, Colin Jenkins, 19, and Trevor Phillips, 20, all from St. Clears YFC, Carmarthenshire, and David Roberts, 23, from Montgomeryshire. Their trainer, Mr D J W Jenkins, said: "The boys have really worked hard. In one week we travelled over 700 miles judging at different Welsh shows."

Individual champion was English team member Geoffrey Smith, of Durham, who scored 336 points out of 400 to win the *Farmer and Stockbreeder* trophy.

Minister's wife takes over

John Winter 17.8.64

Mrs Mary Soames, wife of the Minister of Agriculture, has taken over what was to have been her husband's last big public engagement of the present Parliament. On September 18 Mr Soames, who has a dislocated pelvis, was due to inaugurate the £11 million Great Ouse flood protection scheme.

Now Mrs Soames will pull the levers to open the Lark Head sluice at Barton Mills, Suffolk, and unveil a commemorative plaque on the relief channel head sluice at Denver, Norfolk.

The scheme which has taken ten years to complete, is believed to be the biggest of its kind in Europe. It will completely protect 189,000 acres from flooding and provide benefit for another 251,000 acres.

GENERAL

Land values

John Winter 30.1.64

£162,000 was paid for the 1,066-acre Bayons Manor Agricultural Estate in North Lincolnshire by Mr Edward Sheardown, of Marston, near Grantham, at an auction yesterday.

John Winter 28.2.64

Record price of £292 an acre was paid for Hill Farm, Louth, Lincolnshire, which was sold by auction.

John Winter 28.3.64

£49,000 – £256 an acre – was paid for 191-acre Manor Farm, at Welton, near Lincoln, at a public auction.

Farming's secret men make their move
John Winter 9.4.64

The farmers' "secret service" trading organisation, Agricultural Central Trading, switched its headquarters from London to Chesham, Buckinghamshire, yesterday.

From modern premises equipped with the latest devices, including an electronic computer, the company will mastermind the bulking of orders for farmers' supplies which save producers millions of pounds in discounts from manufacturers. It works through a countrywide network of regional offices and small farmer-groups – but with great secrecy. ACT refuses to publish any of its trading turnover or membership figures.

Chairman Sir Miles Thomas told farmers, suppliers and union officials at the opening of the new headquarters yesterday: "It has been rightly said that one of the ACT's greatest strengths is

that it refuses to declare its strength, and at this stage I see no reason to change that policy."

Although it has been in business only two years, I can reveal that the company has an annual turnover approaching £5 million and a membership of over 10,000 farmers, split up into several hundred small groups.

Sir Miles forecast: "I confidently anticipate our numbers and turnover will be doubled in 1964." He said that the initial expense of preparing the company for business in a big way was over. Farmers would benefit even more in the future from the savings which would be made. The 30-mile move from London would save £20,000 a year alone.

Mr Harold Woolley, NFU president who opened the offices, said: "the ACT's existence has saved British farmers millions of pounds in the last 12 months. It is in fact now geared for an increase of up to 100 per cent in turnover, with probably only a 15 per cent increase in overheads."

Lightweight drainpipes reduce labour
John Winter 20.5.64

Plastic drainpipes, less than one-thirtieth the weight of traditional clay pipes, may soon be criss-crossing the fields of Britain.

A man can carry 200ft of pipe, in 20ft lengths, on his shoulder, and a team of three can dig and lay a drain at the rate of 18 feet a minute.

Plastic pipe costs about the same as clay, but ease of handling is expected to reduce labour costs by half. Schemes using it will qualify for the Government grant of 50 per cent, and its life is estimated at 50 years.

Eight miles of it are being sunk into the Royal Show's new permanent ground at Stoneleigh, Warwickshire, and sections of transparent pipe in inspection pits are being made so that visitors can see them at work.

Australians love a pint of milk

Peter Bullen 20.5.64

The Australian cricket tourists are all heavy milk drinkers. Before they take the field they each knock back at least a pint, sometimes more. Their secret did not come out until their match with Glamorgan at Cardiff this week.

All the Glamorgan players are great milk drinkers, so the Milk Marketing Board decided to put in a milk dispenser to cope with their thirsts. The plan went astray when fitters found there was no point in which to plug the dispenser in the home team's dressing-room, so they put it in the Australians' room. In 2½ days the tourists knocked back 15 gallons of milk; and they want more.

Several players told team manager Ray Steele that the milk was a swell idea and he has asked the Milk Board if they can fit dispensers in all their tour dressing-rooms. This will mean making arrangements for about 20 dispensers to be installed all over the country, but the board is working on the problem.

The South Wales dairy festival organiser, Mr Raymond Jones, was delighted when Mr Steele approached him about supplying more milk. He said: "It will be a tremendous boost to our national dairy festival plans, which this year are linking milk drinking and sport. These Aussie boys can certainly drink it too. Peter Burge told me he thinks it great stuff to

drink, just what was needed when he came off the field."

Patriotically, he added: "Mind you, our Glamorgan boys can match them pint for pint."

The Glamorgan club has such a good milk-drinking record that the Milk Board put up prizes for their Bank Holiday games for the best performances. Gwyn Hughes, the 23-year-old batsman who scored 92, collected £35 for the best innings. Peter Walker won £20 for his bowling and Alan Rees £20 for fielding.

Subscriptions lost *Peter Bullen 22.5.64*

The National Federation of Young Farmers' Clubs is losing subscription money because some branches have a hidden membership which they are not declaring, Mr Ted Newbould, chairman, said yesterday. He told the federation's conference in London that returns this year showed a total membership of 48,486 only, 128 less than last year. But this was not a true return.

"I know there is a hidden membership, and if only we could get a true return then we could show the outside world that our membership is increasing."

Afterwards Mr Newbould explained that some branches were not disclosing their full membership because they would have to pay more to the national federation in subscriptions. He estimated that the federation was losing about £200 a year because of this, and that about 1,000 members could be added if true returns were given.

Mr Newbould suggested during the conference that the federation should forget for ever the idea of changing the name, Young Farmers' Club. A lot of members had been looking for a better name for many years, but could not find one. It has been suggested that some people did not join because the name implied that it was an organisation only for young farmers. But anyone interested in farming, agriculture generally or the countryside can join if they are under 25.

The Earl of Halifax was elected president in succession to Prince Philip, who is retiring after three years.

Holdings merged *John Winter 30.5.64*
The number of farms in Lincolnshire has dropped 15 per cent in ten years through merging of holdings.

Royal Show ready to expand

John Winter 7.7.64

A Royal Show which will set the pattern of farm exhibitions for the next century opens today for the second time on its lovely 200 acres of tree-lined parkland at Stoneleigh Abbey, Warwickshire. The show has improved on its rain-soaked mudbath last year with a big extension of internal roads, enlarged electricity and water system and a spacious new members pavilion.

But the real hard cash programme needed to turn Stoneleigh into the agricultural centre of Europe depends on the vital question of whether the Royal will anchor itself there permanently. By Thursday, when the Royal Agricultural Society Council meets on the ground, members will have a good idea whether the colourful programme they have laid on will attract as many, or more, than the 112,000 who flocked here in mixed weather last year.

Mr Francis Pemberton, honorary director, said: "As soon as we know we are staying here the expansion programme will begin. It will cover a five- to ten-year period of development, with empha-

sis on layout, making the ground as weatherproof as possible, and adapting it for events outside the show period. How much we spend depends on how much we make."

The society is estimated to have saved £50,000 this year in costs of uprooting, transporting and re-erecting the show which faced it annually when it moved round the country. On the spending side it has laid out £12,500 on roads and another £5,500 on other services.

The most revolutionary experiment in the Royal's 125-year history is a lavish late night spectacular in the main ring on Thursday. The programme includes an hour-long pageant, community singing and dancing with Billy J Kramer and The Dakotas.

Hat trick at the Royal Show

John Winter 8.7.64

Yesterday at the Royal Show was hat-trick day for Mr George Gill, 50, a 250-acre farmer at Spillingfleet, near York. His seven-year-old British Friesian bull Hillam Siemkes Adema won the male championship of his breed for the third year in succession.

For Mr Edward Lewis and his son, Leslie, of Dilwyn, Hereford, it was a great finale to their finest year. Their unbeatable pair, Havenfield Eclipse and Havenfield Regina, which have already won the titles of Hereford bull and female of the year, were on top again.

The Wych Cross herd of sleek black Aberdeen Angus beef cattle, owned by Sir Harold Samuel, at Angmering, Sussex, proved the advantage of buying the right breeding stock. The heifer Evaka of Wych Cross, which won the breed title for the second year running, is a daughter of Newhouse Jewlian Eric, the bull

Sir Harold bought for 28,000 guineas.

For working farmer Mr George Dart, from Exmoor, it was a winning climax to his first major season. He won the Devon beef breed championship with his two-year-old heifer Champion Clara 45th.

Sheep entries were down by 62 but there was a fine display from 25 breeds. The record price of 600 gns paid for a Clun Forest ram a few months ago by 48-year-old Mr Sydney Price of Shropshire proved a wise investment. The ram became supreme champion of the breed.

The sound of honey

Peter Bullen 21.8.64

A retired BBC sound engineer has invented an electronic "stethoscope" which, he claims, can double honey crops and cut labour costs. The £40 listening device can detect when a colony of bees is preparing to swarm.

Until now the only way of discovering this has been by inspection every nine days and this involves taking the hive almost to pieces. Labour costs are high, about 1,000 to 2,000 eggs are lost, a lot of honey is wasted and quite often the investigation is fruitless. It can take place up to ten times during the swarming period.

This waste so appalled 62-year-old Mr "Eddie" Woods, of East Molesey, Surrey, that he tried to find ways of avoiding it.

About seven or eight years ago he tried out a listening device and shortly afterwards heard the sweet sound of success. After spending days listening and recording different sounds he discovered that the bees made an unusual bubbling noise before swarming.

For the past few years he has been perfecting the device which is simply attached to the back of the hive by a rubber suction pad. The sounds are measured by a small indicator on top of the device which is about the size of a small book.

After demonstrating it on a hive at the Molesey Apiary Club yesterday Mr Woods said: "I estimate that 90 per cent of the labour costs are saved and the honey crop can be doubled. For a large bee keeper with say 5,000 hives and annual labour costs of £15,000 a year, this would be a big saving on the wage bill alone. It's strange, but Virgil wrote about 'a wailing of the bees' but it was left for me, over 2,000 years later, to identify what the sound meant and to make use of it."

Mr Woods said bee keepers were anxious to prevent swarming because, unless they caught a swarm, they lost a valuable contingent from their stock. A swarm weighing four or five pounds contains about 20,000 bees worth £4 or £5.

1965

Farmers' fury over their treatment during the annual price review dominated the early months on the political front. The workers' union called for state-owned farms and the industry was to get its "Little Neddy" in December. Fred Peart's "Golden Pitchfork" was supposed to ease the less profitable out of the industry. The workers lost their claim for a 40-hour week but did get a record pay rise of 8 shillings (40 pence) a week! The record harvest was at risk from the appalling weather and the livestock sector saw the import of more French cattle.

 POLITICS and FOOD

Farmers fight for more cash

John Winter 10.2.65

Farmers' leaders go to the Ministry of Agriculture today to open the most critical annual price review talks for years. They are pledged to fight for the second instalment of their demand for a 25 per cent increase in income over three years, but the prospect of repeating last year's achievement of a £31 million increase in the total value of their guarantees is not bright.

Mr Fred Peart, Minister of Agriculture, is anxious to live up to his promise to maintain the economic health of the industry, but he is under strong Cabinet pressure to keep down the cost of living. He has proposals, which will be welcomed by the farmers' unions, to put more money into the pockets of hard-hit small farmers, especially those on hill land.

He wants to encourage greater production of commodities which would replace imports and reduce the trade gap, and he also wants to extend the effects of the review beyond the scope of one year, and build in long-term planning.

The price increase Mr Peart can allow for milk will be the most crucial single decision in the review. An increase of 3*d* a gallon would mean ½*d* rise on the present retail price of 9*d* a pint, or, as is more likely, £25 million a year in subsidy.

Farm fury on squeeze

John Winter 18.3.65

Farmers' leaders are to ask the Prime Minister for an immediate meeting to protest about yesterday's farm price review.

They are angry because they were awarded only £10.5 million extra in the total value of their guarantee, instead of the £30 million they wanted.

The main cause of the fury is the rise of only 1*d* a gallon on the price paid to producers for milk. The farmers had asked for 6*d* and confidently expected 3*d*. However, their 1*d* increase means that the price of milk to the housewife will go up by ½*d* to 9½*d* a pint from August.

The farmers' disappointment last night brought threats of unofficial action. In Sussex, Mr Arthur Philipson, chairman of the 65-strong Mayfield branch of the National Farmers' Union, was planning to use tractors to cause gigantic traffic jams during the Easter and Whitsun Bank Holidays. At an emergency meeting of the branch yesterday the farmers voted to block the London-to-Eastbourne and London-to-Brighton roads.

Mr Philipson said: "We shall drive our tractors up and down these roads at a snail's pace on Easter Monday and Whit Monday. We shall be out for three hours, one in the morning, one in the afternoon and one at night. I am hoping to get 100 farmers to take part, some who are not even in the union at the moment, and I hope this protest will be taken up nationally."

A spokesman for the AA said: "This plan would cause chaos on the roads, especially to Brighton, which is used by about 3,000 cars an hour at holiday time."

The Mayfield men also plan to cause a big milk shortage to the public. Instead it would be manufactured for powdered skimmed purposes. They will also burn and tear up all Government correspondence sent to them.

Other dairy farmers in 17 counties are so angry with the Government and the National Farmers' Union that they are planning a breakaway body to represent them.

Their leader, 36-year-old Mr Alan Woodruff, of High Holden, Kent, said: "We are thinking of some concerted national protest about this award." It would be on the lines of a token withdrawal of supplies – about 10 per cent of milk output – which they could keep up for a long time.

Mr Woodruff said he was planning a meeting in London soon of representatives from the 17 counties to form a national dairy farmers' association to negotiate with the Government. He claimed that hundreds of farmers in the United Kingdom had said they would support the new body.

Mr G T Williams, deputy president of the NFU, said: "We will not do anything to encourage unconstitutional action, but we cannot answer for moves by our branches. It may well be that we have a revolution on our hands." Mr Williams led the farmers' review team in negotiations with the Government, which had to impose the settlement when the farmers rejected it – for the sixth time in ten years.

Farm costs have gone up by more than £28 million in the past year, so the settlement, which was announced in the Commons yesterday by Mr Fred Peart, the Minister of Agriculture, means that the farmers will have to find an extra £18 million out of their own pockets. The 1*d* a gallon rise in the present guaranteed milk price is barely enough to cover the increased cost of wages, National Insurance, Bank Rate, petrol tax, etc., which have piled up in the past year.

The total value of the increases is exactly one-third of what the farmers got from the Tories last year, and one-third of what they hoped for this year. As a result, other guaranteed prices are down.

This table shows the main changes:

Commodity	Price	Change
Beef cattle	174s a live cwt	+4s
Pigs	44s 9d a score deadweight	−1s 7d
Hen eggs	3s 8.36d a dozen	−1d
Milk	3s 5.85d a gallon	+1d
Wheat	25s 5d a cwt	−1s 1d
Barley	25s 4d a cwt	−1s 4d
Potatoes	285s a ton	+5s
Sugar beet	130s 6d a ton	+2s 6d
Sheep and lambs		
Duck eggs		
Wool	No change	
Oats		
Rye		

There are promises to help farmers to become better businessmen, with grants for management records, and strengthen their marketing position with credits for co-operative and group trading. But the most significant long-term plan is a hint – the first to come from Whitehall – of measures to encourage the voluntary amalgamation of uneconomic farms into viable and well-equipped holdings.

In the Commons, a roar of outrage came from farming-area Tories when the review was announced. Eric Sewell, Daily Mail *Parliamentary Correspondent, reports*:

> Tories shouted "Resign!" at Mr Peart. Several times, more than a score of indignant Tory MPs were on their feet seeking to complain. Sir Martin Redmayne, Opposition spokesman on agriculture, said the Minister's plans would be greeted with dismay. He asked Mr Peart bluntly: "What are your intentions – do you intend to resign?"
>
> But Mr Peart clearly showed he intended to do no such thing. He claimed that the award to milk producers was

reasonable, right and fair. What was more, apart from last year – an election year – it was the most generous award ever made to milk producers.

Where the money goes

Peter Bullen 18.3.65

The Government has full control over the price of milk. It controls the price the housewife pays, the amount the producer receives, and the amount the dairy company can take as profit and to meet its costs.

Out of the 9d the housewife is paying today for a pint of ordinary milk, the dairy company takes about 3d, made up of 2¾d for bottling, processing and distribution costs, and ¼d profit.

On paper the farmer gets about 6d, but because a lot of the milk he produces is sold for making into cheese, butter and cream, at lower prices, over his whole range of production it works out at 4½d a pint. After taking away his production cost the farmer is left with about ½d a pint profit.

Farmers making a great mistake, says MP
Peter Bullen 20.3.65

The farmers' revolt against the Government's price review gathered fury yesterday with more threats of militant action.

But last night the Government hit back at the critics. Mr John Mackie, Parliamentary Secretary to the Minister of Agriculture, said at Laurencekirk, Kincardineshire: "This review settlement is perfectly fair. I believe farmers are making a great mistake by protesting so violently. This is simply not justified by a calm look at the facts."

There had been far tougher settlements under Tory Ministers, he said and there had not been anything like the fuss. Mr Mackie added: "Farmers must be careful. They will lose the public sympathy built up over the years if they act irresponsibly." The Government wanted to see higher farm incomes, "but not simply by increased hand-outs from the taxpayer and the ordinary consumer."

Earlier, the National Farmers' Union said 17 branches had held protest meetings about the review. Ten passed a vote of no confidence in the Minister, Mr Fred Peart, and seven called specifically for his resignation.

Farmers start 'tell the people' campaign   John Winter 26.3.65

Full page advertisements in national newspapers have been booked for Sunday and Monday in an unprecedented effort to win public support for the farmers in their fight against the Government's price review settlement. They will cost £37,000, and will be the first shot in a "fair play for the farmer" campaign for which the National Farmers' Union has asked its 190,000 members to subscribe £500,000 at the rate of £2 minimum or as much as they can afford.

The Farm Price Campaign group was in continuous session in London yesterday, sifting suggestions which have poured in from 59 branches. Even before the fighting fund was launched, cash was pouring in. At a mass meeting at the Corn Exchange, Dorchester, 1,500 farmers threw more than £2,500 into the hat in a few minutes. It was the biggest farmers' rally in the history of the county NFU branch, whose total membership is only 3,100. Individual subscriptions have come in from sympathisers outside agriculture.

The NFU said: "The advertisements are only the start of a big build-up. We are working on other publicity ideas, including a 500,000-circulation campaign newspaper. It will be of four tabloid-size pages and will be circulated to every farmer next week, with copies for distribution wherever they can be used advantageously. There will be stickers for farmers' cars, farm vehicles and those used by suppliers. We are also considering including leaflets in a big range of pre-packed foods which farmers supply to housewives." The union has never attempted any comparable campaign in its 50 years' history.

The newspaper advertisements set out the facts of the farmers' position, as the NFU sees them. They stress the difference between the average income rise of the community of 56 per cent, in real terms, in 12 years, and the 1 per cent received by farmers; and the rise of 3.2 per cent a year in national productivity, compared with 6 per cent in agriculture.

A special train has been booked to bring 1,200 farmers from Devon, Dorset, Somerset and Wiltshire to London for the agriculture debate in the Commons next Wednesday.

No cash boost yet for farmers
John Winter 8.5.65

Prospects of securing any improvement in farmers' incomes before next year's Price Review are negligible. This became clear last night, after discussions lasting all day between farmers' leaders and the Ministry of Agriculture, following the Downing Street dinner on Thursday.

A statement from No 10 said Mr Wilson made it clear that the review could not be reopened. "On the other hand," it added, "the Government would be glad to examine the new proposals for future policy, which had been put forward, and to consider their attitude to the implications."

Teams led by Sir Harold Woolley, NFU president, and Sir John Winnifrith, Permanent Secretary at the Ministry, argued all day yesterday over a programme for talks on the farmers' nine-point plan. But last night they broke up, without even agreeing on a joint statement on the proposals. The talks will be resumed on Monday.

The outcome of the Downing Street dinner and yesterday's talks will be reported to an emergency meeting of the 144-strong NFU council on Tuesday, but it is apparent that the new negotiations cannot produce any early cash boost for farmers.

Comment – A story of success

[This item appeared as the leader in the London edition – Editors]   20.5.65

What a bore that phrase about "the mess the Tories left behind" has become.

Ministers, not realising that it is an excuse for their own failures, repeat it, parrot-like, day after day until the rest of us feel like screaming. Mr Fred Peart, the Minister of Agriculture, was at it again yesterday, twice, in the Commons. Speaking of increasing the efficiency of farming, he said: "I am sorry I have a backlog of 14 years of neglect."

Now let us take hold of that foolish phrase and examine it in the light of what has really happened. Let us see if Mr Peart can in any way be justified. British agriculture is one of the most efficient of modern times. It is the most highly mechanised in Europe and probably the world. The Russians take it as a model of what they would like their own vast industry to become. Before the war we produced one-third of the food we con-

sumed. Today the proportion is one-half, though the population has increased by 10 per cent.

British farming is one of the great success stories of the post-war era. The achievement is partly rooted in tradition, partly the consequence of the war, and partly the result of measures introduced after the war. The most important was the Agriculture Act of 1947, the basis of latter-day agricultural economics. It was a good Act, introduced by a good Labour Minister of Agriculture, Mr Tom Williams. Credit where credit is due. But since he left office phenomenal progress has been made. Tory Governments passed ten Acts for the improvement and prosperity of the land between 1951 and 1964.

Here are a few farming figures: Number of cattle: 1951, 10.3 million; 1964, 11,620,000. Sheep: 1951, 19.9 million; 1964, 29,650,000. Pigs: 1951, 3.9 million; 1964, 7,739,000. Poultry: 1951, 94 million; 1964, 112 million. It is the same story with beef, milk, eggs, mutton. Potatoes and oats have decreased but most other arable and root crops, wheat, barley, sugar-beet and so on, show large increases.

Average farm earnings have risen from £6 5s in 1951 to £12 5s today and farm incomes from £296 million to £475 million.

Critics may complain that the Conservatives paid the farmers too much. They may prefer State buying of commodities to private trading. They may disagree with other aspects of their policy. But to label the fair picture they left behind "Neglect" is to make a fool of words and to show that those who make the accusation do not know what they are talking about.

The same is true of the other activities of Tory Governments, and perhaps our few farm statistics may help to prove the point.

Call for State-owned farms

Peter Bullen 18.6.65

If Britain's 400,000 farm workers had their way the Government would take over all agricultural land, and every farmer would become a State tenant. This land nationalisation is the feature of the National Union of Agricultural Workers' first policy statement in 12 years, published this morning.

The general secretary, Lord Collison, said yesterday: "More than 300,000 workers have left agriculture since the war, but very few farmers have. Many of our units are far too small to be viable. There has to be rationalisation, and some farmers have got to go. Farmers have got to accept that there has got to be more co-operation between them."

The policy statement says: "A rational distribution of farm units and the provision of adequate fixed capital investment will be obtained only through the public ownership of land. The Minister of Agriculture should have power to take over areas of land large enough to be managed as an estate or a number of estates, and to place them in the hands of a land commission.

Immediate reaction of the National Farmers' Union to the land nationalisation suggestion was to dismiss it curtly. The NFU president, Sir Harold Woolley, said: "My members will treat it with the contempt it deserves."

Having disposed of farm land to the State, the NUAW would then like to see public ownership of the big companies which manufacture the three main groups of requisites farmers need: machinery and fuel, fertilisers and feeding stuffs.

Pay-off – Peart offers golden pitchfork

Peter Bullen 18.11.65

The Government introduced a Bill yesterday to modernise agriculture and improve meat marketing. It provides for millions of pounds to be spent on a "golden pitchfork" plan of gratuities and pensions for small, uneconomic farmers who can be persuaded to stop work on the land.

It provides an extra £80 million for farm improvements over the next few years for farmers who stay. And it allows for the setting-up of a Meat and Livestock Commission – six or seven men sharing £50,000 in salaries – to improve the production, marketing and quality of meat supplies.

The Minister of Agriculture, Mr Fred Peart, announced in a White Paper in August his plans to give golden handshakes to farmers who leave so that their land can be amalgamated with other farms in larger, more economic units. Actual pensions and gratuities have not been decided yet. They may be even higher than those proposed in August after protests by farmers that the amounts are too small.

Farmers who take over vacated land will be eligible for grants of up to half the cost of the amalgamation work (but not the land) providing the new holding will give full-time employment for the farmer and at least one farmworker.

The cost of the pensions, gratuities and amalgamation grants will run at £15 to £17 million in a few years. About the same amount a year will be spent by the Government on grants for farm and hill land improvements, improvements in farm business recording, agricultural and horticultural co-operation incentives, and the establishment of rural development boards in the problem hill areas. The boards will have wide powers to control sales of land and to license private afforestation and will be able to buy land themselves and give financial assistance for amenity improvement schemes.

The £80 million for farm improvements announced yesterday is an extension of the present Farm Improvement Scheme, which gives one-third grants for approved projects. The rate of grant is being cut to one-quarter, but as the range of eligible improvements is being extended, the scheme is still expected to cost £11 million a year.

Mr Peart's proposed Meat Commission will be advised by committees of farmers, distributors and consumers. But it will not be able to buy or sell meat or have any control over imports. Farmers argue this will make it ineffective. Butchers claim that it will put up the price of meat, as it will be financed by a levy on all animals slaughtered in Britain. Details of the levy scheme to provide an estimated £2–£3 million a year have not been worked out yet.

Now farming gets a little Neddy   Peter Bullen 23.12.65

A little Neddy for agriculture is to be set up after months of discussion between the Government and farming organisations. It will be the 20th little Neddy to be formed to deal with individual industries under the wing of the National Economic Development Council.

The organisation will be concerned with how far the Government intends to let the industry expand under the National Plan. It will determine how much British produce can be substituted for imports.

Membership of the committee is still being worked out and is not likely to be announced until the end of January. As well as six management and four trades union representatives, the committee will have an independent chairman, a member from the Department of Economic Affairs, Neddy, the Ministry of Agriculture, and possibly an expert from outside the industry.

The purpose of the economic committees, or little Neddys, is to provide a forum where both management and workers' organisations can sort out together problems affecting the growth, efficiency and productivity of their industry.

At the same time, they can maintain a constant two-way exchange of information and views with the Government through the Government representatives sitting on the committees.

CROPS and MACHINERY

1,500 learn safety with chemicals   John Winter 19.1.65

More than 1,500 farmers and farmworkers have applied for specialised training in the safe use of agricultural chemicals. This astonishing response to an offer of half-day training courses in main arable areas has been received by

Fisons, manufacturers of farm chemicals. The first course was held at Holbeach, Lincolnshire. More than 60 others have been arranged for the next two months.

The scheme was announced last September by Lord Netherthorpe, Fisons' chairman, who was president of the NFU for 15 years. It followed a pilot scheme last year and was, he said, designed to raise the standard of application of spray chemicals and minimise any risks involved in their application.

A Fisons' spokesman, involved in organising the courses, said that in many cases the number of farmworkers expected to attend had been almost doubled by the number of farmers wishing to accompany them. Most of the courses are organised through agricultural merchants, and will be held on their premises. There is a short talk, a specially made film, and practical demonstrations.

● On December 30, Harry Longmuir, who conducted a *Daily Mail* investigation into toxic farm sprays, reported the Rural Councils Association as saying: "There is no public confidence in the safety arrangements, which are largely unknown, even to farmers."

Letter of the week – responsible chemical use

W Broughton 1.2.65

I think it is time that the public was told that farmers use chemicals to kill weeds and pests – not crops.

Farmers do not run around with tins of poison in their pockets. Some at least try to keep abreast of new developments. I first attended a lecture on the use of chemical sprays in 1952 and the infor-

mation has been reviewed periodically since them.

The robot to lift us out of the 'horse and cart' age

John Winter 12.2.65

A machine that can load itself in the field and then take the load to a store at 20 mph and stack it will be tested this summer. This revolution in produce, fertiliser and feed handling was reported to the National Power Farming Conference at Brighton yesterday.

Mr John Holt, research engineer at the National Institute of Agricultural Engineering, Silsoe, Bedfordshire, told 450 farmers that their tractors and trailers are only a "poor development of the horse and cart." The tractor was not suitable for load carrying, only pulling. But on most farms men and machines spent more than half their time handling materials, so the institute had thought up the vehicle specifically to carry a load. It could load a number of units on itself with a fork-lift device, carry them at reasonable speed, then unload and stack them. It made material-handling a one-man, one-machine operation.

One firm is building a prototype and another is negotiating a licence to build.

Still on robots: The Milk Marketing Board creamery at Llangefni, Anglesey, is to install a computer next month to keep daily details of consignments from its 950 suppliers.

Machine blazes neat new trail through weeds   *John Winter 10.3.65*

Two agricultural supply firms yesterday launched a fiery crusade through Britain's countryside, with a new technique to get rid of weeds by burning.

Flame cultivation is claimed by Shell-Mex and BP Gases and Maywick Appliances to destroy weeds and unwanted foliage at about half the cost of chemical sprays. It will have an instant appeal to the opponents of sprays, because it depends only on intense heat in the right place at the right time.

A new machine to do this job with the use of propane gas, and which goes on sale today, is the brainchild of Mr Ron Mayes, 55-year-old farm appliance maker, who started his career in his family's century-old firm of agricultural ironmongers at Wickford, Essex. It has previously been best known to farmers for poultry appliances, and this is its first field machine.

Four years ago Mr Mayes built his first hand-made "burning-up" machine. He said: "People thought I was crazy. But I was studying the development of flame cultivation in America, where it was being used on an entirely different pattern of crops, and I was sure it could be used on our farms. We are now miles ahead of any other country in Europe in this technique, and I believe it has a big export potential."

Already a French company has negotiated to produce Mr Mayes' flame cultivator on licence. The machine, towed by a tractor, will direct naked flame at varying heights and intensities, with such precision that it can be used between rows of plants in a growing crop, without damaging them.

Used to burn off potato haulm and weeds, before harvesting, it will do the job at 35s an acre, against about £3 for chemical methods.

A representative of Shell-Mex and BP Gases said: "Flame cultivation is an additional weapon in the farmer's arsenal in the increasing war on weeds." A machine with 12 burners on an 8ft boom costs £336.

Poison down on the farm
John Winter 13.4.65

Fears of widespread poisoning of wild life, farm animals and even people by the the use of chemicals in agriculture will be examined in a two-day investigation which starts this morning. Some of the best scientific brains have been working for months on evidence which will be submitted to a conference at the Yorkshire Institute of Agriculture, Askham Bryan, near York.

The chairmen of seven study groups, which have investigated different aspects of the problem, include university scientists, directors of experimental stations and a medical consultant. Their studies have ranged from the general use of chemicals in farming to their application and effects in forestry, horticulture, crops, wild life and domestic animals. The possible dangers of chemical farming, which have caused most public concern, will be left to the final session tomorrow.

The whole conference will be summed up by Sir Harold Sanders, former scientific adviser to the Ministry of Agriculture, who is probably the best informed man in Britain on the subject.

Tractors boost exports
John Winter 28.4.65

Export orders worth £700,000 were announced yesterday when Ford's £20 million factory was officially opened at Basildon, Essex. The factory will produce £40 million worth of exports a year from the new plant, said Mr Jim Bywater, executive director. This represented 1 per cent of Britain's entire export earnings.

The new orders are for 100 tractors for Morocco, 100 for Iraq and 500 for Colombia, South America.

Between 300 and 400 tractors a day will soon be coming off the 1¼-mile production line at Basildon, 75 per cent for export. It is the most modern tractor factory in the world. It will supply parts to two factories in America and Belgium and 26 assembly plants in countries all round the world.

Safety plan for killer tractors

Peter Bullen 26.6.65

Mr Fred Peart, Minister of Agriculture, is planning steps to cut the high death-toll caused by overturning tractors. He said yesterday that new laws making safety frames or cabs compulsory for tractors were now being given "urgent consideration".

Mr Peart was speaking at the opening of a new machinery testing unit near Penrith, Cumberland.

Overturning tractors consistently cause the highest number of farm deaths each year, although the total has been falling recently. Of the 106 farm victims last year about a third died in this way. Twelve people have died because of tractor accidents already this year. And the worst accident months are still to come.

Unions were pleased to hear of Mr Peart's concern. A National Union of Agricultural Workers' official said: "We have been pressing this issue for some considerable time."

Millions lost in harvest disaster

Peter Bullen 17.9.65

The state of Britain's corn harvest is getting very serious, Sir Harold Woolley, National Farmers' Union president, said yesterday. Millions of pounds have been lost by the farming community and the public. He added: "Over the country as a whole I estimate that only about 50 per cent of the corn is in, and of the rest 20 per cent has been flattened by storms."

Yields were down and the cost of getting the harvest in was much higher than usual because of the difficulties caused by the weather. From all over the country NFU observers have sent in gloomy pictures. In some places in the North-East only 30 per cent of the crops have been harvested. Other areas of Yorkshire, Lancashire and Cheshire have got in more than 40 per cent.

From the Midlands there are indications that the harvest will be the worst since the war. Thousands of acres of corn have been submerged under green undergrowth which is slowing down the combines. The acreage in the Midlands is up by 77,000 acres, but the total amount of grain harvested is not likely to be any more than last year.

In the South-West farmers are rapidly reaching the stage where the rest of the harvest is becoming a salvage operation. Both quality and yields are deteriorating rapidly.

In Wales less than half of the corn has been gathered. Heavy rain has caused considerable flooding. In many upland areas of North Wales farmers are still struggling with the remnants of the hay harvest and vast quantities have had to be written off as worthless for winter feeding. A little more progress has been made in the South-East as farmers snatch what corn they can from the fields between showers.

More syndicates   *John Winter 17.11.65*

First machinery syndicate was formed in Hampshire ten years ago; now 113 other groups have sprung up in the county and own £217,185-worth of machinery.

About 230 of the county's 3,000 farmers belong to syndicates. The idea has spread all over Britain and has led to the setting up of more than 600 other syndicates.

Enter the Mini tractor for the small farmer   *Daily Mail Reporter 1.12.65*

A Mini tractor is announced by the British Motor Corporation today. There is a range of mini-accessories to go with it, and there are hopes of a big export appeal in agricultural areas.

The tractor was designed to provide small farmers, horticulturalists and industrial concerns in Britain with a cheap form of mechanisation. It is only 8ft 2in long and 3ft 8in high. The conventional tractor is more than ten feet long.

But the biggest reduction is in horsepower – 15 compared with between 40 and 65 hp. The power unit, a new four-cylinder 948 cc diesel engine, the smallest in volume produced in this country, is based on the BMC's 948 cc Mini car petrol engine. It has a nine-speed gearbox, including three reverse gears, and a top speed of 13½ mph.

But first reactions by home farmers were not too hopeful.

Mr Peter Boulden, chairman of the Kent branch of the National Farmers' Union machinery committee, said he did not think the Mini-tractor would have any appeal for small farmers. "Look at it this way," he said. "Most of them now are buying second hand tractors sold off by the larger farmers because they can't afford the capital outlay on a new tractor. I cannot see them buying this machine. Still it could well have a large market among root crop growers, market gardeners and people like that. It may well be an ideal machine for them."

The standard model costs £512, and the more refined model, with front loader equipment, hydraulic power unit, and other extras, £585. This is more than £100 cheaper than a basic medium-size conventional tractor. The biggest tractor costs about £890.

 # LIVESTOCK and POULTRY

Breeders want no more French cattle   *Peter Bullen 20.1.65*

Britain's leading pedigree cattle breeders are alarmed at suggestions that the Minister of Agriculture, Mr Fred Peart, is thinking of allowing more French Charollais cattle to be imported. They say it could jeopardise our £750,000 a year export trade in pedigree cattle. Foreign buyers who are impressed with Britain's disease-free cattle would shop elsewhere as they would fear the imports would bring diseases from France.

Mr William Turpitt, secretary of the National Cattle Breeders' Association, said yesterday that the association had always opposed the Charollais, first because they were not superior to British breeds and secondly because of the concern over the effect on exports. The Livestock Export Council is writing to the Minister today to voice its concern about the danger to exports.

Mr Cyril Manning, secretary of the South Devon cattle society, said the Ministry tests showed South Devons to be superior to Charollais in crossing trials with dairy breeds.

Rustlers have price on their heads   *John Winter 26.2.65*

Rustlers of farm livestock have a price on their heads in Berkshire. Rewards of up to £50 for conviction and recovery of stolen animals are being offered by the county NFU. Rustling is common in several areas and other branches are expected to follow.

Cattle, sheep and pigs have disappeared, and the raiders have even stolen fertilisers stacked in plastic bags in fields. The number of thefts has doubled in recent weeks: one farmer has been raided three times. The reward is 10 per cent of the stock or material recovered, up to a maximum of £50, provided the rustlers are caught and convicted. Most of the raids have been in the downland areas in the west of the county.

In other counties, raiders have concentrated on open country remote from farmhouses, like Dartmoor, and the dales of Derbyshire and the North. But last week a Herefordshire farmer had 23 fat sheep stolen at night from a flock of 146 in a field at Croft, near Leominster. They were worth £160.

The NFU parliamentary committee secretary, Mr Monty Keen, said: "Reports we are receiving indicate careful organisation, involving a number of operators, equipped with livestock transport. Thefts on this scale can cost a farmer his whole profit on a batch of livestock in a few hours."

Animal entitlements
Peter Bullen 27.2.65

A plea for an "animals' charter" is made by the Soil Association in its evidence to the Government inquiry into factory farming.

The association says: "While animals are carrying out the service of producing food for us they are entitled to a good and healthy life."

Rabbits by the million for the table   *Peter Bullen 2.3.65*

Europe's biggest turkey producer, Mr Bernard Matthews, of Norfolk, has started a ten-year plan for Britain to eat twice as much rabbit meat. He is supplying a network of 20 large breeders with a special fast-growing hybrid rabbit. They in turn have started supplying parent stock to dozens of small rearers, who will produce about 4 million table rabbits in the next two years.

Mr Brian Daws, who manages the Matthews rabbit enterprise, said: "At present we eat about ten million rabbits a year in Britain. We hope that within seven to ten years consumption will rise to 20 million."

Behind the white hybrid rabbit is a six-year history of breeding experiments which has resulted in a four-way cross. Two female lines have given the rabbit good mothering qualities, plus size and frequency of litters, while the two male lines have contributed to its ability to put on meat quickly and economically.

Mr Daws said: "The young rabbits reach market weight of 4lb within eight weeks. Producers are getting about 1s 8d to 1s 10d a lb for them at present. A 4lb live rabbit produces two 1lb pre-packs for the housewife to buy at about 5s 6d a lb."

He said that the company had begun to meet many orders from abroad.

Farmer v Egg Board fight today

John Winter 23.3.65

Mr James Sperling, the Suffolk egg farmer who tried to get the Egg Marketing Board "arrested" in January, is himself a "defendant" in a board disciplinary court today. He and three other producers will be accused in London of contravening board regulations in disposing of eggs from his 2,000 layers at Brandeston. He believes it may be the court to end all board courts.

He said yesterday: "I have been seeking to have this case brought for nine months."

The case involves two charges: selling eggs other than to the board, and selling unstamped eggs, not directly to consumers. The charges are made against a company of which Mr Sperling is a director. He will be represented by a barrister and will challenge the right of the board to hold the court, and even to exist. Also present will be farmers from other areas, who have joined Mr Sperling in collecting signatures for a petition to demand a poll which could abolish the board.

He said: "There are now over 2,000 signatures, twice the number required. I do not want to lodge the petition until the Minister of Agriculture reveals his plans for changing egg marketing procedure, but it may be difficult to dissuade some of our supporters from taking immediate action."

Farmer James tames Little Lion

John Winter 24.3.65

Farmer James Sperling yesterday smashed the grip of the Egg Marketing Board on the nation's egg sales. A charge of breaking the "Little Lion" rules by selling unstamped eggs was dismissed by one of the board's own courts.

Afterwards Mr Sperling said: "We have broken the board's monopoly. It is a limited victory." And he made an offer to other egg producers: "We are prepared to give advice absolutely free to anyone who wants to know how to do it."

Board leaders now have to devise new rules to plug the Sperling loophole before too many of their 9,000 million eggs a year slip through it.

Mr Sperling, with 6,000 layers at Brandeston, Suffolk, is to continue his "Little Lion-taming" plan, which gives him, on average, nearly 1s a dozen more for his eggs. When all his birds are laying this will mean a margin of about £100 a week and 300 dozen large eggs a day will go out from the farm.

Part of his plan was described yesterday. The hens are owned by one company, of which he is controlling director. This company, Hillcorse, is registered with the board and sells about half the total egg output direct to consumers, which is allowed. The other half are given away – not sold – to another company, Yes Laboratories, of which Mr Sperling is also director, though the majority shares are held by Hillcorse. Yes Laboratories sells the eggs to shops, where they are sold, unstamped, to housewives. The board has powers only over producers and is unable to act against Yes Laboratories.

Under an earlier plan devised by Mr Sperling retailers were given eggs free on condition that they bought, from Yes Laboratories, booklets worth about 2d but costing 12s 6d. This scheme was dropped in favour of the later, more efficient, one.

Mr Sperling still has one secret – the method by which one company can give eggs to the other without going bankrupt. But he is willing to share it with farmers who want to join him in beating the board.

Long wait to buy the big French cattle
John Winter 26.4.65

The biggest sales drive for a foreign breed of cattle to be mounted on the British market has been launched by French breeders of the massive Charollais beef animals. Cashing in on the approval given by Mr Fred Peart, Minister of Agriculture, for female as well as male animals to be imported, a syndicate headed by M Emile Maurice, president of the International Association of Charollais Breeders, is buying full-page advertisements in the farming press.

British breeders are ready to pay a stiff price for the nucleus of a Charollais herd, but their hopes of getting them this year are slim. Strict quarantine rules, coupled with limited facilities on both sides of the Channel, will rule out any customers whose names were not down before Mr Peart gave his approval on March 11.

This year's intake looks like being limited to 230 animals, the capacity of the French quarantine station at Brest where they must stay a month before entering Britain. The British Charollais Society had applications for more than that number before Mr Peart's all-clear. So far they have failed to find quarantine quarters which will meet Ministry standards in this country, where the animals must be held for another month before they go to their buyers' farms.

Letters of the week – identifying cows
3.5.65

Readers are full of ideas for fixing up cows with "rear number plates", to help cowmen identify each animal in the milking parlour.

All suggestions are acknowledged here, and have been forwarded to the Milk Marketing Board, which is offering a prize of £500 for a satisfactory method.

Here are some of our readers' ideas:

- Adhesive medical tape, stencilled, and attached to the rump. *John Clements, London, E7.*
- Bone, or plastic bangles placed round the legs. *A J S Brown, Bishops Sutton, Hampshire.*
- An identity tab fitted round the lower part of the tail. *S Ramsden, Hamstreet, Kent.*
- A nylon stretch stocking on one leg. *C Marples, Ashford in the Water, near Bakewell, Derbyshire.*
- A mirror, fixed at the head of the stall, and focused so the cowman can read through it the identification number wherever it is carried on the cow's body. *W M Cummings, Birmingham.*
- A shaped identity plate to stand up above the tail head, and held by a strap with tail loop, which would extend along the back and connect with a neck loop, carrying a similar plate. *J L Wood, Holme Lacy, Herefordshire.*

Britain buys bulls from Canada
Peter Bullen 7.5.65

A consignment of very special young bulls will arrive in England from Canada three weeks today. They are 15 Canadian Holstein bulls – all under a year old – which the Milk Marketing Board has bought for more than £6,000. Purpose of the importation is to see if the bulls can improve the milk and butterfat production of English dairy cows and whether they can improve the fleshing qualities of young stock.

The board's selection team, which returned this week, visited 53 farms and chose the 15 from the pick of Canada's pedigree herds. They will all be going to the board's artificial insemination centres and the first four will be in use by late autumn this year. The others will become available between November and next April.

Letter of the week – time to stand and stare
Ian Wimberley, Newquay, Cornwall 10.5.65

What has happened to stockmen that they need assistance in identifying their charges? Some shepherds know every ewe in the flock individually. When I kept poultry I knew most of my 150 hens and could tell if any were missing. Cattle are much easier to identify. I have no trouble with my Friesians.

Modern milking methods leave no time for stockmen to stand and stare. A few minutes each day just looking at the cattle would be a lot cheaper than complicated indexing.

EMPLOYMENT

Farm men lose 40-hour week claim
Peter Bullen 4.5.65

A claim for a five-hour cut in the 45-hour standard working week of Britain's 325,000 farm labourers was turned down yesterday.

The Agricultural Wages Board, which had heard the workers' and the employers' cases at previous meetings, spent three hours over the claim in London before rejecting it on the combined votes of the employers' representatives and the independent members.

Lord Collison, general secretary of the National Union of Agricultural Workers, who led the workers' side, said he was surprised and disappointed at the outcome. "We think it is the strongest claim we have ever put for a reduction of hours. Since our last reduction in March 1963 more than 100 industries have had their hours cut."

He said that 300,000 workers had left agriculture since the war at the rate of 20,000 to 30,000 a year, a fall of nearly 50 per cent in the labour force. He added: "This decision will speed up the movement from the land."

Mr Henry Sharpley, National Farmers' Union chairman, said that because the workers had had a 6¼ per cent wage increase as recently as January (12s a week) and because the five-hour cut would have cost £37 million, the employers had no alternative but to oppose the claim.

Farmers plan to share labour force
John Winter 22.6.65

Three farmers in Yorkshire are planning a shared labour force with their workers living together in a hostel. It is the first time this has been attempted in Britain.

Labour pools, which are forecast as the answer to farm work problems, would overcome the difficulties of ensuring continuity of skilled work, for which a specialist is paid high wages, and of providing good living standards on a commercial basis in remote areas.

The three farmers who are exploring the idea are Mr John Stringer, of Bishops Wilton, and two brothers, Mr Robert and Mr R B Sleightholme, of Youlthorpe. Their plan developed out of a move by the Pocklington NFU branch to experiment with a shared apprentice scheme. This proposed that a young worker, in training under the farm apprenticeship scheme, should work for more than one farmer, but the idea was turned down by the East Riding NFU executive. Mr Stringer said yesterday that many points about the scheme were still unsettled.

He said: "Our main aim is to get workers of high calibre, and also to meet the problem of housing, which is difficult in this area. We realise that not all workers would welcome the idea of living in a hostel, but we think it would appeal to some, and especially to young men just out of college."

The basis on which workers would be engaged is not yet determined, but it is likely that the farmers will form a syndicate for this purpose. Mr Stringer is an arable farmer. The Sleightholme brothers concentrate on milk and pigs.

What's this? A farmer's boy collective! John Winter 8.9.65

A big family house in a good-class residential district of Pocklington, on the edge of the Yorkshire Wolds, is about to become a milestone in the history of the farmer's boy. From next week it will be the home of young workers in the first shared labour force ever organised by farmers to meet the new demands of the technical revolution in agriculture.

Eight farmers will share the labour of six workers, who will live communally, with a full-time housekeeper to cook, clean and wash for them, and make the house, Hayton Villa, a home-from-home. The new workers will not displace any men employed individually by the farmers. The aim is to train them into a highly mobile force of technicians, who can be employed either singly or collectively by any member of the employers' syndicate.

Because this roving farm work is obviously a young man's world, and initially a single man's, the syndicate's first two recruits are 16-year-olds. They are both town boys, one from Hull and the other from Bradford, with a burning desire to make their life on the land. The farmers will engage one more boy of the same age, and fill the remaining places with older workers, who have been through an agricultural college or institute. The youngsters will receive basic training on farms and their employers hope this will encourage them to take a college course.

Why, I asked the farmers, go to the trouble and expense (and the cost is formidable) of setting up a syndicate to employ, house, feed and mother their workers? It is a very different proposition from shared tractors, or combines run by the machinery syndicate.

The answer lies partly in the type of farms and country and partly in the changes the farmers foresee in the pattern of farm work and employer-worker relations. The eight farms are all in the medium size range – smallest 72 acres, biggest 340 acres – and have one to four full-time workers in addition to the farmer and his family. They are all highly productive and efficiently run, but good farm cottages for married workers are few. All the farmhouses, which are within a radius of eight miles of Hayton Villa, date from the time when workers lived-in, but life on a remote farmstead on the sparsely populated Wolds has not much to offer a modern boy.

"Both farmers and workers have come to regard living-in as a thing of the past, which should not be revived," said Mr John Stringer, a founder of the syndicate. "We hope that with the accommodation and services we are providing, the boys will enjoy their home life as much as their work. They will have the amenities of a lively little town right here on their doorstep."

The syndicate's farms offer all kinds of arable crops, dairying, pig and sheep production, and, initially, each worker will divide his time between two farms. Mr Stringer, with 260 arable acres and pigs, at Bishop Wilton, is to share one with Mr Edwin Stephenson, who has a 40–50 cow dairy and grass farm at Melbourne. "We shall keep the arrangements as flexible as possible for the benefit of both sides," said Mr Stringer.

I was sorry to learn that so far the Agricultural Workers' Union has not been in contact with the syndicate, for if it succeeds other farmers will undoubtedly want to copy it. The union is sceptical about dual employment, but it is interested in the prospect that could result in better pay for a specialist, by keeping him fully employed on his particular job.

The capital and imagination put into the scheme by the farmers, and their high standard of efficiency, make it clear that they are not candidates for Mr Fred Peart's amalgamation scheme for worn-out farms and farmers.

Young farmers rebel at milking drudgery
John Winter 19.10.65

Young farmers are refusing to follow their fathers into a life of seven-days-a-week drudgery in the cowshed.

This is clearly shown by the Milk Marketing Board in an analysis of the £1 million-a-day dairy farming industry in England and Wales. It shows that more than one-fifth of dairy farmers are over 60, more than half over 50, and only 4 per cent under 30. The figures are from a sample of 5,000 of the 108,500 farmers who were producing milk in 1963–1964. By last month the total had dropped to 98,600.

Why are young men fighting shy of dairy farming? New entrants in the past, suggests the report, have been predominantly farmers' sons, but the aspirations of the present generation have changed.

Dairy farmers under 30 years old in most areas have smaller herds than those between 30 and 50, but in the 50s and 60s there is a marked decline in the size of the herds. "It seems reasonable to suppose that the large investment of capital, necessary for expansion on modern lines, is unlikely to be undertaken by those approaching retirement."

Other reasons stand out:

- *One:* More than two-thirds of all dairy farmers have a total turnover of under £5,000 a year, and only 12 per cent exceed £10,000.
- *Two:* Only 11 per cent of farms are equipped with modern labour-saving milking parlours. On 86 per cent the job is done in cowsheds.
- *Three:* Half the farmers employ no hired help.
- *Four:* Combined effects of inadequate labour and milking resources means

that on average one man milks 19 cows at 12 an hour. In the biggest herds a man averages 47 cows at 21 an hour, but in the smallest herds a man takes more than an hour to milk six.

All this is hardly surprising since more than 70,000 cows – 3 per cent of the national herd – are still milked by hand.

8 shillings for farm workers

John Winter 21.10.65

Farmworkers were awarded their biggest rise since the war yesterday – a pay increase of 8*s*, and a reduction of one hour in their working week. It brings their minimum wage to ten guineas for 44 hours.

Total value of the award, by the Agricultural Wages Board, is estimated at 15*s* a week for the 350,000 adult wholetime workers in England and Wales.

Neither the farmers' nor the workers' union expect any criticism from Mr George Brown, the Economics Minister, or from the Prices and Incomes Board.

The award was also being agreed unanimously by the board after negotiation between the farmers' and workers' representatives. Mr G T (Bill) Williams, deputy president of the National Farmers' Union, said farmers wanted to do their best for their workers within their capacity to pay. "This increase was negotiated against the background of a very bad farm price review, but the movement in industrial wages generally has been such that we have to keep a reasonable parity for the agricultural worker."

It is estimated that the award will raise average earnings on the farm from £13 16*s* 7*d* a week to about £14 11*s*, compared with the £18 18*s* 2*d* average in industry as a whole.

Lord Collison, general secretary of the Agricultural Workers' Union and immediate past chairman of the TUC, welcomed the settlement. He said: "I am satisfied it is the best we could get at the present time." Both he and Mr Williams believe the award will put a brake on the drift of workers from the land to higher paid jobs in other industries, but Lord Collison said the position could never be satisfactory until the farmworkers achieved complete equality.

Bumpkins? Not us, say men down on the farm

John Winter 18.11.65

Farmworkers have had enough of the cartoon character with straw in his ears and his trousers hoisted with string. They want a new image, in keeping with their role as "skilled, sophisticated men working with a complex of tools and machinery."

The National Union of Agricultural Workers is being pressed by Berkshire members to present this new picture through a publicity campaign. Mr Charles Brown, the union's district organiser in Berkshire, said yesterday: "My members are thoroughly sick of being shown as yokels. They have to be fully skilled mechanics and impromptu veterinary surgeons. They work an average 52-hour week – and their poor public image is reflected in a ridiculously low wage of only 4*s* 6*d* an hour."

PEOPLE in the NEWS

Rebel back at work

John Winter 25.2.65

Rebel farmer Jack Merricks walked happily across a potato field at Lodge-lands Farm, near Rye, Sussex, yesterday – a free man.

Earlier yesterday he had been released from Brixton Jail, where he still had six days to serve on a 14-day sentence for refusing to pay a £50 levy to the Potato Marketing Board. The £50 was paid, without his knowledge, on Tuesday, by another potato farmer, Mr Richard Hayward, of Great Bromley, Essex, who was concerned at the effect the imprisonment might have on the image of the Board, which he supports.

He said: "This is our Board, and I thought it was time its supporters had something to say."

The order for Mr Merrick's release reached the jail by express post yesterday. He said: "When I was told to get my things together, I thought I was being moved to another cell." Instead he was discharged and taken to the main gate where his wife, Margie, and son, Peter, were waiting.

"I am glad to be out of jail, but I did not want to get out this way," he said. "I do not know Mr Hayward, and I think it is a great pity that if he has got money to give away that he should give it to the Potato Board, instead of a more deserving body. I only hope he will be equally generous with other potato growers who are less able to pay their levies than I am, and I advise them to send their demand notes to him."

Mr Merricks drove straight home and spent the rest of the day on two of his farms, where potato planting was in full swing.

He now plans a meeting with Mr Gordon Charity, of Deeping St James, Lincolnshire, a grower who launched a campaign against the board while he was in jail. Last night Mr Charity said he had 1,000 signatures supporting his campaign to abolish the Potato Board.

540 Years of Service

John Winter 26.4.65

Twelve Devon farm workers who between them have 540 years of service in the county are to be given long-service awards at Devon County Show at Exeter next month. Top award goes to 70-year-old Mr Neilius Soper, who has worked on the same farm at Powderham for 54 years. Mr John Elson, 62, and Mr George Farley, 67, who worked on the same farm as Mr Soper for more than 40 years, will also get awards.

Its England's cup after 12 years

John Winter 8.7.65

Four farmers' sons scored a great triumph for England at the Royal Show when they won the Young Farmers' International Cattle Judging Contest. They were the first English team for 12 years to win the *Daily Mail* Gold Cup. It was presented to the captain, Michael Saxby, 23, of Collingham, Nottinghamshire, by Princess Alexandra.

The other members of the team were Tony Watson, 23, of Collingham, Nottinghamshire; John Leeming, 24, of Carnforth, Lancashire, and Brian Clarke, 21, of Clitheroe, Lancashire.

England beat Ulster by 32 points, then came Scotland, Wales, Isle of Man, New

Zealand, Australia, Kansas (US) and Jamaica.

The Princess presented each team with Royal Agricultural Society medals. The competitor with the highest individual score, Brian Clarke, received the Farmer and Stockbreeder Trophy, and each competitor got a commemorative medal from the *Daily Mail*.

Who follows Sir Harold at the top? *John Winter 20.12.65*

Sir Harold Woolley's announcement that he will retire from leadership of the National Farmers' Union next April has started speculation and recrimination.

Sir Harold's denial that his decision was influenced by health considerations was reported exclusively in Saturday's Farm Mail. It has been interpreted as meaning that it was the result of persistent criticism of his leadership by a minority of NFU members.

This, then, has roused widespread indignation among his supporters, who admire the selfless way in which he flung himself into the union's "fair deal" campaign after a two-month illness, even though they may not agree whole-heartedly with the policy planning behind it. Critics of his tactics in combating the Government's unpopular price review settlement last March have demanded more militancy – without indicating how it could be effectively applied in such a widely dispersed industry.

Another issue on which Sir Harold has been criticised is his handling of the union's mounting financial difficulties, which led to a demand for heavy pruning of 1966 estimates by the treasurer, Mr Fred Vincent. A debate on this at last week's council meeting led to a narrow defeat for the president on a proposal that the union should withdraw financial support from the Agricultural Central Co-operative Association, to which it has paid £54,000 at £6,000 a year. This brought fears of re-opening the old war between the co-operative movement and the NFU, which most farmers would prefer to see working closely together.

But the council's decision to refer the issue back was apparently not treated as a vote of confidence, either by Sir Harold or by those who voted against him. When, at the close, he said he would not seek re-election, he made it clear that he had contemplated retirement for some time.

The union now faces an embarrassing 15 weeks of manoeuvre and intrigue over the succession to the presidency, which 60-year-old Sir Harold has held for six years. In the next few weeks his supporters will make strenuous efforts to persuade him to change his mind. There is no doubt that if he consented to stand for a seventh year, he would win a majority in the 144-strong council. But he is unlikely to relent, and speculation has already started on candidates for the job, worth about £7,000 a year.

Those certain to be nominated are Mr G T "Bill" Williams, the deputy president, who stood in for Sir Harold during his illness, and Mr Henry Plumb, the vice-president. A third probable starter is Colonel "Jock" Wilson, deputy president before Mr Williams, and a permanent member of the council.

One or two outsiders may be put up by individual counties, and it is even possible that the list may include names not at present on the council. The rules limit nominations to council delegates, who are elected by county branches throughout England and Wales.

GENERAL

£1m plan for Royal Show

John Winter 6.7.65

The Royal Show, which opens its four-day run at Stoneleigh Abbey, Warwickshire, today will be developed into the £1 million agricultural centre of Europe within ten years.

As columns of wagons carrying the elite of Britain's livestock and hundreds of tons of machinery, farm supplies, food and drink rolled into the 152-acre ground yesterday, Lord Netherthorpe, Royal Agricultural Society president, unfolded the plan. He said: "Our object is to fulfil a bigger role on a much more permanent basis by creating a focal point where every section of our great industry can co-operate, to promote the practical advances of British agriculture."

Since the Royal settled at Stoneleigh two years ago, £840,000 has been spent on buildings and services. Next week, work starts on a £12,000 scheme to drain and resurface the main ring and build up viewing embankments.

Princess Alexandra and the Duchess of Gloucester will fly in and out of the show by helicopter. Princess Alexandra will visit the pavilion of the Young Farmers' Clubs and present the *Daily Mail* Gold Cup to the winning team in the international cattle judging contest.

Royal lands £357,000 US order

John Winter 9.7.65

The export-chasing Royal Show landed a £357,000 order for British farm tractors from America yesterday. This plum follows hundreds of other orders and inquiries which amply justify the effort to make the show the agricultural industry's premier export shop window.

Mr Francis Pemberton, honorary director of the show, said: "This is far and away the best Royal for exports we have ever had. We have never really gone out for exports on this scale before, and I would expect the business which has been started here to amount to many millions of pounds."

The Duchess of Gloucester, who flew in and out of the ground at Stoneleigh Abbey, Warwickshire, by helicopter, spent six hours on a detailed tour.

A record influx of nearly 3,000 overseas visitors in the first three days was headed by 500 buyers. They included a party of Russians, who have been at Stoneleigh each day. France ordered £20,000-worth of maize- and rice-drying equipment from E H Bentall, of Maldon, Essex, and Whitlock Brothers, of Halstead, Essex, are negotiating a Turkish order for sugar-cane trailers and fertiliser distributors.

Many hundred pedigree animals of breeds which have made Britain the stud farm of the world will be exported as a result of visits to the show's livestock lines by overseas breeders. The biggest single consignment will be £30,000-worth of Galloway beef cattle, which will be shipped from Hull to Russia on July 21. Another key shipment completed at the show was for 14 heifers and two bulls of the choice beef-producing Aberdeen Angus breed which will go to Ceylon.

Spey Cast, a five-year-old brown gelding, owned by Mr W F Ransom, of Wellingore, Lincolnshire, crowned a brilliant show season by winning the Hunter Championship yesterday. He was ridden by Mrs Ransom. The Duke of Edinburgh's trophy for all-round stockmanship produced a thrilling one-point victory for 50-year-old Scot, Mr Alexander Forsyth, who manages Mr Ian Hamilton's 800-acre farm at Melrose, Roxburghshire.

Twenty-one farm workers were presented with long-service awards by the Duchess of Gloucester.

Coo-cooo-cuk . . . it's the dove that's disturbing the peace

Peter Bullen 27.11.65

There is trouble brewing in the suburbs. Immigrants from the Far East are finding life so comfortable on the fashionable fringes of the country's towns that their numbers have doubled every year since the early 1950s. Local inhabitants, forced out by the new arrivals, are setting up home in the country or in less favourable residential areas of town.

The immigrants are officially named Streptopelia decaocto – otherwise known as collared doves. Before 1952 they were unknown in Britain. Today there is hardly a county in the whole country where they have not been spotted. In 1963 the population of collared doves was estimated at just over 10,000. Last year the figure was 19,000 and by now it is probably 30,000.

At first they were spied with great delight by ornithologists, but nowadays they are almost commonplace. Some authorities are already forecasting the collared dove will rapidly become a major pest and that it is time it was named officially as a pest so that it could be destroyed without infringement of the wild bird protection laws.

Government ornithologists working for the Ministry of Agriculture are not convinced that it is such a serious threat. The Ministry's pests department has just completed a countrywide check on complaints from farmers, landowners and gamekeepers about the dove and there is little evidence that it is being a nuisance.

Perhaps the main reason is that the collared dove dislikes living in the depths of the country. He prefers nesting in trees in people's gardens, preferably pines, around the fringes of villages and towns where he can be fairly easily recognised by his call, a slightly monotonous coo-cooo-cuk. He likes to eat lazily, by raiding chicken runs or game bird sites where grain has been scattered, freshly sown fields, or around corn stores, mills or any place corn might be spilled.

There is little evidence that the collared dove will give up his easy way of life to live in the country where in winter particularly he would have to switch his diet and at the same time compete with birds like the wood pigeon for the small supplies available. It seems he has found a niche for himself as a suburbanite though he has ousted other species of pigeons from our back gardens to make enough room for his family.

1966

This was a time of national economic crisis, with growing inflation, a wages freeze and a credit squeeze. Against this background, the Minister of Agriculture, Fred Peart, warned the industry not to disrupt the food supply, and Prime Minister Harold Wilson intervened in the Price Review settlement. Agronomists appeared on the scene, foot and mouth disease fears were voiced and the eradication of brucellosis from the national herd was talked about. The establishment of the Agricultural Training Board started a long battle with the industry over how it should be funded. After hesitating to stand for re-election, Sir Harold Woolley finally lost the Presidency of the National Farmers' Union.

 POLITICS and FOOD

Now milk will come in plastic bags
Peter Bullen 14.1.66

The Milk Marketing Board announced yesterday a sales drive of milk packed in plastic bags fitted into cardboard boxes.

At first the packs will be only in the three- and five-gallon sizes for use by caterers and in vending machines. Future uses planned by the board include one-gallon packs for people to use at home, said the board chairman, Mr Richard Trehane.

This would bring two advantages. It would help to maintain milk supplies in homes where regular daily deliveries are impossible or too expensive, and the use of larger containers than pint bottles could lead to more milk being drunk. A board official said: "It has been proved that where there is greater availability of milk, consumption is also greater." In America, where wives had milk on tap in similar types of packs, sales in homes had increased about 10 per cent.

Farmers to be fined for drugged milk
John Winter 21.1.66

Farmers who sell milk containing penicillin or other antibiotics after April 1 will be "fined" 1s a gallon. The Milk Marketing Board penalty scheme will apply to all farmers who ignore a warning.

Contamination follows the treatment of cows for mastitis, a disease of the udder. Usually penicillin cream is injected. Farmers are asked not to sell milk from treated cows for 48 to 96 hours, according to the drug use. The milk can be used for feeding calves.

A Government investigation in 1963 showed that 14 per cent of all milk in England and Wales had traces of antibiotics, and the Ministry of Agriculture asked the Milk Board for urgent action.

It has taken two years to find a testing formula which can be applied by all dairies and will not confuse contamination from antibiotics with reaction to dairy sterilising agents. Last October antibiotics were still found in 2.2 per cent of milk tested in England and Wales.

The Milk Board will also introduce in April regulations to encourage farmers to use bulk storage tanks for milk and do away with old-fashioned churns. Milk is piped straight from the cows into tanks in farm dairies, from which it is collected by road tankers and delivered straight into bigger tanks at collecting depots.

To attract small farmers into the scheme the Board proposes they should use tanks big enough to hold two days' milk, which the tanker will empty on alternate days. As an inducement, farmers are offered loans of up to 75 per cent of tank cost, repayable over three years. Throughout this period they will get a premium on their milk which will repay loan and interest. Grants of 25 per cent for farm improvements, proposed under the Agriculture Bill, will cover the balance of the cost.

Peart warns farmers

John Winter 26.1.66

Imports will be used to make up food shortages if farmers cut supplies in protest against an unsatisfactory price review next month. Mr Fred Peart, Minister of Agriculture, gave this warning in a 45-minute meeting yesterday with Sir Harold Woolley, President of the National Farmers' Union. He asked to see Sir Harold about a five-point questionnaire, circulated to the union's 182,000 members, on possible militant action if next month's review was not satisfactory.

A Ministry statement later said that Mr Peart "made it plain that he regretted a step which could only be construed by public opinion as a threat, issued before discussions had begun, of industrial action if the farmers' wishes were not met. The Minister emphasised that if an artificial shortage of food were created it would be his responsibility to ensure sufficient food supplies regardless of their source."

Sir Harold dismissed the meeting as a "bit of a storm in a teacup". But he insisted that issuing the questionnaire was "a perfectly reasonable and normal procedure for an organisation seeking to ascertain its members' views on a matter of vital interest." He added: "I have nothing to withdraw or apologise for. The NFU is not inciting a strike, or seeking to threaten the Government. We are threatening nobody."

The vital question which upset Mr Peart was: "Would you be prepared to support action to reduce supplies of a commodity on the basis that a substantial majority of producers of that commodity would do likewise?"

Sir Harold agreed that if action was taken on this it would lead to "some reduction of supplies. That does not mean we would be deliberately creating shortage or holding the public to ransom."

The questionnaire went out a fortnight ago when union representatives were asked to deliver it personally to every member. The first batches of replies show a substantial majority in favour of action. In some areas, fewer than 10 per cent of farmers said they would not support it.

The tension which has developed in the past 48 hours may affect Sir Harold's personal future. Tomorrow he will tell the National Council whether he will stand by his decision not to seek re-election in April, or bow to pleas from many members to stand again. Some members, who criticised him for inadequate action after the unpopular 1965 review, are now wondering whether to end their opposition to another term of office. They believe that his withdrawal now would weaken the farmers' prospects in the review, which starts on February 9.

Letters of the week – Peart's "threat"
John Winter 31.1.66

Mr Fred Peart, Minister of Agriculture, upset many farmers by describing the National Farmers' Union questionnaire about militant action as "a threat".

His own threat, to bring in imported food to make up any shortage caused by farmers' industrial action, added to their resentment. These are some of their views:

N L Earle, Warminster, Wiltshire

Who does Peart think he is to start waving a big stick? His minions cut more than £150 from my income last year to try to prevent increasing British grain output from damming the flow of low-quality French grain into this country – grain which could not be paid for by exports and which carried twice as much subsidy as ours.

J M Harvey, Mayfield, Sussex

Since Mr Peart has uttered a threat to farmers, he should now assure the NFU before negotiations begin that he can negotiate and agree a Price Review, and will not be countermanded by the Prime Minister as at the last Review.

Mrs Margaret Couling, Marlborough, Wiltshire

What Mr Peart and those in authority do not seem to be able to get into their heads is this: every other employee in the country either gives warning or intention to strike, or just down tools to get what he wants (and generally does get it).

All the NFU is trying to press home is the fact that farmers must be recompensed sufficiently in order to pay their men – who they look on as their friends and fellow workers – their rightful, living wage.

Can they survive on heartbreak farm?
John Winter 21.2.66

The following words, from a letter sent to me by a Somerset woman farmer, a widow, should be tape-recorded and played over before every session of the annual Farm Price Review, now being argued in Whitehall between farmers' leaders and the Government: "He had to leave and take a job elsewhere, while I struggle to keep my son's inheritance going somehow, so that one day he can take over the farm."

The widow, Mrs Joan Pile, knows that her future, that of her son and the farm will be decided by the result of the review, expected on March 16. It will not only fix prices for a year but show farmers how the Government intends to interpret the guidelines for agriculture laid down in the National Economic Plan for the next five years.

At least 70,000 other farmers will learn their fate from Mr Fred Peart, Minister of Agriculture, that day. I fear that if not a death sentence, he will give no more than a limited reprieve.

Lue Farm, Winsham, has been in the family of Mrs Pile's late husband for almost 100 years. Its 140 acres cover one of those narrow, secluded valleys in Somerset, whose sides are so steep that most of the fields are useless for anything but livestock farming. In Somerset those conditions spell milk, as they do in Devon, Cornwall, Wales and the Pennines. In these areas the threatened small farmers are thick on the ground, and the price of milk is again the crunch of the price review.

After the war the profits from dairying, poultry and pigs provided a comfortable living for Mrs Pile, her son, Patrick, just out of agricultural college, and her invalid daughter. Mother and son worked hard and they employed one full-time farm worker. They built up a herd of 50 Ayrshire cows, installed a 1,000-bird egg-laying unit, and a piggery to turn out 150 porkers a year. In 15 years they invested £8,000. When Patrick married and started a family he moved to a comfortable, modernised house down the lane. Mrs Pile bought a home in the village for the farmworker.

Then the tide turned. In the past five years costs of labour, feeding stuffs, fertilisers, machinery and other requisites, as well as electricity charges and rates, have shot up, while the rise in income from farm products, screwed down at successive price reviews, has lagged far behind. Three years ago Patrick, then 30, left home to become manager of a big farm in Hertfordshire.

His mother, now past 60, still starts work at 6.30 am. But her personal income is considerably less than the £15 a week she pays her two farm workers. One of these is about to leave because she cannot afford to pay him more. The farm overdraft is mounting.

Similar stories of human tragedy and heartbreaking, unrewarded devotion to the land could be repeated from tens of thousands of British farms. Bewildered small farmers, who have never known any other way of life, bitterly accuse successive Governments of betraying them. Perhaps they are right, because we will still depend on farms of under 150 acres for more than half our milk.

Some are rounding on their own union leaders and setting up "rebel" organisations, like the 20,000-strong National Dairy Producers' Association.

The National Farmers' Union faces a bitter dilemma. It maintains that the farmers' case in the price review can be fought effectively only if the claims of all are presented together. At best it can win only a compromise, and it cannot conceal the fact that big farmers blessed with ample capital for expansion and intensification are making reasonable profits out of prices which are ruining little men.

Mr Peart's dilemma is worse. He is fixing prices backed by Exchequer guarantee. If he raises them to a level which would restore prosperity to the heartbreak farms, he would pour money into the pockets of wealthy farmers. Apart from inviting an embarrassing surplus of production, which would depress market prices, he would send subsidies soaring. But that is no excuse for abandoning small farmers to bankruptcy.

To his credit, Mr Peart has recognised their plight in his "Golden Pitchfork" scheme. This offers grants or pensions to encourage small farmers to give up hopeless holdings so they can be amalgamated into larger, more economic, units. But he will have to improve on the niggardly £500 plus £15-an-acre grant, or £100 plus £1-an-acre pension, and do something for those who are determined to continue on their farms.

A phrase in last year's price review

White Paper about "releasing resources for use elsewhere in the nation's economy" is one of the worst examples of unintentional callousness ever to come out of Whitehall. Can Mr Peart visualise a bred-in-the-bone Somerset farmer of 50 (and most of the hard cases are older) leaving his fields and village and going to work happily and effectively in a Bristol plastics factory?

No. The Government must make it possible, for those who are willing and able, to continue at least to normal retiring age, at a decent standard of living. The inheritance which Patrick's mother is struggling to preserve is part of the inheritance of us all. Can we afford to lose it?

How Wilson stepped in to aid farmers – Farm Mail special on the Price Review   *John Winter 17.3.66*

The Government gave farmers an extra £23 million in price guarantees yesterday. This leaves the farmers to pay £9 million of their increased costs out of higher efficiency.

Soon after details of the award were announced, it was revealed that the Prime Minister had taken a big part in the price review. This was to make sure a settlement was reached that the farmers' unions would accept and try to "sell" to their members. Details of Mr Wilson's intervention were given by Sir Harold Woolley, president of the National Farmers' Union.

Sir Harold said: "I had two talks with the Prime Minister, and also a telephone conversation. The review talks were getting rather near explosion point when we first met. His attitude was understanding and helpful, especially on the question of the Government's long-term assurances,

which we consider so important, and which are included in the review White Paper."

Sir Harold described the review as "just, but only just". He went on: "We would have liked more money, but, as responsible citizens, we recognise the seriousness of the nation's economic position."

Improved prices paid to farmers through the review would not mean any significant rise in food prices, said Sir Harold. Milk will go up about ⅛d a pint (charged at ½d a pint for the first four months of next year). Other increases are about 2d a lb on beef, ¾d a lb on lamb and mutton, and 1/37d a lb on potatoes. There are cuts in farmers' prices for eggs and wool.

Main criticism among farmers will centre on an increase of 1d a gallon for milk, which will leave dairy farmers hardly any better off. The increase will stir resentment among customers, who will have to pay an extra halfpenny a pint for four months, beginning next January. This will raise the price of ordinary milk to 10d a pint for those months.

The Government presented the review as a first stage of its National Plan to enable farmers to embark on a long-term programme of expansion. Favourable changes are centred on beef, starting with a direct increase of 10s a cwt on fat cattle, and a new subsidy of £6 10s for cows kept to rear calves. To keep up other meat supplies and cushion a continuing shortage of beef, fat sheep and lambs get a price rise of three farthings a pound, and pigs for pork and bacon are raised by 9d a score (20lb).

But over milk, the most vital product related to beef, the review leaves an ominous question mark. It gives farmers an extra ½d a gallon on the guaranteed price, and raises the number of gallons on which it is paid by an extra 50 million. Although the National Plan recognised that more beef must mean more milk (because most of it comes from calves born to dairy cows), farmers will conclude that the Government is not prepared to pay a reasonable price for the milk.

The breakaway National Dairy Producers' Association yesterday issued a statement condemning the review. It accused the National Farmers' Union and Milk Marketing Board of false optimism and calling for measures to cut milk production. It said the review spelled out the Government's intention to drive some 50,000 dairy farmers out of business.

"This milk award is the biggest confidence trick ever," said Mr Alan Woodruff, NDPA chairman.

The new guaranteed prices are:

		Rise or fall on last year
Beef cattle	184s a live cwt	+10s
Sheep	3s 2.75d a lb carcass weight	+¾d
Milk	3s 6.35d a gallon	+½d
Potatoes	290s a ton	+5s
Wool	4s 5.25d a lb	−2d
Hen eggs	3s 6.76d a dozen	−1.6d

Mr Joseph Godber, Tory spokesman on agriculture, attacked the review as being inadequate on beef and milk. On the increase for beef, he said he was doubtful whether it was sufficient to be an incentive for higher production.

The Liberals launched an even more scathing attack. Mr Clifford Selly, chairman of the Party's agricultural panel, said: "An important chance has been fumbled. After the promises in the National Plan farmers were waiting anxiously for the green light to expand home food production and save imports."

Farmers applaud Heath on Europe
John Winter 28.4.66

Mr Edward Heath told farmers yesterday that entry into the European Common Market would present them with great opportunities.

An audience which included some of Britain's leading farmers applauded loudly when Mr Heath told the Royal Association of British Dairy Farmers: "The British farming industry is more efficient than that of any country in Europe. Let us grasp this opportunity with both hands, and not advance towards it timorously and apprehensively." He had never looked on British agriculture as an industry which had to be cosseted and protected. Britain could be one of the strongest partners in the Common Market and farming one of our strongest elements.

Mr Heath said that if the present Government support system for agriculture were changed to the Tory plan of import levies it would free the farmer from the charge of being subsidised. It would end a situation where the Treasury was always breathing down the farmer's neck, looking at the cost of support entirely from the point of view of the Treasury balance sheet.

He said that rising demand for dairy produce in the Far East would absorb an increasing amount of the output of traditional suppliers to our market, like New Zealand.

Come off it – We want our pinta daily Mr Davies!
John Winter 21.7.66

Imagine the Postmaster-General saying

that because postmen's wages and other costs are likely to go up he is proposing to deliver our letters only on Mondays, Wednesdays and Fridays. Yet that is the only reason, advanced in an official report, for suggesting that two years from now milk deliveries should be made only on alternate days.

The recommendation was the one-man inspiration of the one man who ought to know better – Mr Jim Davies, former general manager of the Milk Marketing Board. It ought to be called "Mr Davies in Dairyland", but Mr Fred Peart, Minister of Agriculture, is considering it. Two reasons alone should lead him to reject the plan:

1. The majority of housewives are satisfied with the daily delivery system and do not wish to see it changed. This was proved by the Berkhamsted experiment, in which a leading multiple dairy tried to "educate" its customers into going without Sunday deliveries. It was abandoned. Yet Mr Davies believes that housewives could be "educated" into receiving milk only on alternate days.

   Daily deliveries are the backbone of the present high level of consumption – five pints a week for every man, woman and child. This is one of the highest in Western Europe, and has been rising while consumption in other countries has declined.

2. Introduction of long-keeping milk, which Mr Davies suggests would tide us over the milkless days, is a non-starter. Failure to popularise homogenised milk showed that. Housewives want pintas which look like milk, as well as tasting like it. This long-keeping stuff lacks the cream-line by which the average consumer judges whether a bottle is up to standard, and it costs $11d$ a pint, $1\frac{1}{2}d$ more than the standard pasteurised grade and as much as the highest Channel Island grade.

Mr Davies suggests that farmers who sell direct to the housewife should be allowed to continue daily deliveries. There are nearly 10,000 such farmers and together they sell about 4½ per cent of all delivered milk, often in competition with multiple dairy firms. Can you imagine United Dairies or the Co-op agreeing to alternate-day delivery while Farmer Giles is happily putting bottles on his customers' doorsteps every day?

Then there is the suggestion that delivery on alternate days will cut delivery costs from $10d$ a gallon to $5d$. On Mr Davies's own estimates it seems that about 55,000 milkmen are engaged in placing 18,000 tons of milk a day on our doorsteps (not to mention 7,500 tons of glass bottles). In the Davies Dairyland half that number of milkmen are going to hump twice their present load to double the number of houses, and collect twice as much cash each weekend. Come off it, Mr Davies!

Finally, there is the notion that because milk cannot be kept for 48 hours in summer without refrigeration, dairy companies should either rent fridges to customers or help them to buy them, presumably on hire purchase. I should think that Mr Callaghan, the Chancellor, would be intrigued by this one, since 58 per cent of households are without fridges, enough to send the national hire purchase bill through the roof.

Freshness of any farm product, especially a highly perishable one like milk, is a quality the average consumer cherishes and is prepared to pay for. Hence the popularity of farm-gate eggs and the famous farmers' shop at Uckfield, Sussex.

Let us, by all means, achieve saving by increasing the efficiency of the daily milk service. But let us preserve something which everyone likes, which stimulates milk sales to the benefit of farmer, dairyman and consumer, and is the envy of every other country in the world.

Letters of the week – for and against daily milk delivery

John Winter 1.8.66

The idea of changing the traditional daily delivery of milk bottles on the nation's doorsteps has brought many letters from readers.

Here is a selection of their views on John Winter's article on the Davies Report, which recommends milk delivery on alternate days, and suggests long-keeping milk as a useful standby.

J Harvey, Beckenham, Kent

Many housewives would be prepared to accept delivery on alternate days, if they could be certain of the milk's freshness. It is delivered in many places as late as 4 pm, and on sunny days the bottles are so warm the cream has burst out of the top.

(Miss) N Bunce, Birmingham, 31

I have a refrigerator and manage quite well with milk delivered once or twice a week. I have also tried long-keeping milk and cream and found them extremely palatable. If an experiment in reducing the frequency of deliveries receives sufficient publicity, I cannot see why it should not succeed.

(Mrs) Alice McCarthy (Address supplied)

From long experience of the dairy trade I recall the time when milkmen made three deliveries a day, except on Sunday, when they made one. At the end of each day they made up their books, and washed and polished their cans and churns – a 12-hour day, for which their wage was 18*s* a week, plus a small commission.

I am 83, in good health, and I want my pinta delivered daily.

W E D Bell, Chairman, Express Dairy Company, London

I consider the article by your agricultural correspondent misleading. In an experiment at Berkhamsted we reduced deliveries from seven to six days a week and, at the end of seven weeks, were selling more milk than before. Market research in the Greater London area showed that 85 per cent would be happy to accept alternate day deliveries.

Long-keeping milk is fresh milk indistinguishable in flavour from ordinary pasteurised homogenised milk, and will keep for six months without refrigeration. Since we did a sales promotion on it, our sales in the home market are equal to those for export.

(Mrs) M Prickett, Watford, Hertfordshire

My husband and I want daily milk delivery. We have tried long-keeping milk and consider it has an artificial tang. My milkman agrees with you and says he is not going to hump twice his present load.

(Mrs) M K Love, Brentwood, Essex

I agree with all John Winter's objections to alternate day delivery, and can add another. I do part-time work and leave home before the roundsman calls. What would happen to a double supply of milk left on the step all day, perhaps in hot sun?

K Bowden (dairyman) Camborne, Cornwall
In most streets throughout the country several dairymen each have customers who are all supplied with the same milk. If this duplication of transport and labour could be cut out the trade would maintain seven-day delivery for a long time to come.

Farmers warn 25 shillings on your food bill if Britain joins the Six
John Winter 16.11.66

The weekly food bill for a family of four would rise by 25s a week if Britain entered the Common Market with its existing agricultural policy, according to the three United Kingdom farmers' unions. Their estimate, based on a total increase of £700 million (about 12 per cent) was published yesterday. They set out their case for a complete revision of EEC farm policy to meet the needs of a bigger community, embracing Britain, and perhaps Ireland, Denmark and some other European Free Trade Association countries.

The demand for a new look at farming arrangements anticipates findings of a comprehensive analysis of the effects on British agriculture of a link-up with the Six (on existing terms), on which the NFU was working when the Prime Minister announced his new move towards Europe last week.

British farmers would suffer financially, the report says, from the abandonment of guaranteed prices and long-term assurances and production policy which they now enjoy. The biggest shock to

their economy would come from a substantial rise in the cost of grain for feeding livestock. This would go up by about £10 a ton (roughly 50 per cent), bringing an increase of at least £100 million a year in the cost of purchased feed. This in turn would raise the cost of milk 3d to 4d a gallon, pigmeat 6d a lb (deadweight), eggs 6d a dozen, table chickens 3d a lb (liveweight) and beef 10s a cwt (liveweight).

Livestock products represent 64 per cent of total farm sales, and the union contends that although some selling prices would rise, the increases would not balance the rise in feed costs. Market gardeners would be jeopardised by removal of their tariff protection, and the end of production grants for hill cattle, sheep and other commodities would seriously affect their producers.

The NFU recognises that grain and beef farmers would benefit by higher prices at first, and there may also be openings for increased farm exports. These advantages would not be enough to offset the bad effects, and in the threatened sectors all farmers, not just the little men, would suffer.

It emphasises, however, that given what it describes as a sound framework and fair terms, it has no doubt of the British farmer's ability to meet any competition in an enlarged community. "It would be miraculous if the present policy, designed for the Six, were automatically to suit the needs of a wider grouping," it says.

If Britain's approach goes ahead it clearly expects our negotiators to fight for an annual price review, which, it suggests, would benefit all countries in the community.

Up and up goes the bill

Peter Bullen 16.12.66

Prices rose more in the past month than at any time since the freeze began in July. This was indicated by the fourth Price-Meter survey organised by the *Daily Mail* and the Government-backed consumer council.

Since mid-November the average cost of the Price-Meter shopping basket of 37 food and household items has risen by 1*s* 2*d* (1.3 per cent) to £4 15*s* 9*d*. The main cause was the rise in egg prices. Only 20 of the 203 shops checked had not raised prices. The average price for a dozen standard eggs rose by 1*s* or 25 per cent. Eggs went up because the winter fall in output cut supplies. After Christmas they should become cheaper again but other foods, including bread and flour, will cost more.

Price rises for tomatoes (up 5*d* a lb since August) and apples (up 1½*d* a lb) contributed to the extra cost of the shopping basket. The shopping basket continues to cost considerably more in the North and Scotland – although some foods are cheaper – than in the South.

London: Shopping basket cost £4 14*s* 6¾*d*. Up 1*s* 4½*d*. Meat was cheaper in London than in any other region. The average price of a 2½lb joint was 11*s* 5¼*d*, compared with 14*s* 6¼*d* in the North and Scotland.

South: £4 14*s*. Up 1*s* 4*d*. The cheapest region for chicken, butter and pork sausages. A 3lb broiler chicken averaging 9*s* 11¾*d* was up to 6*d* cheaper than in other regions.

Midlands: £4 15*s* 8*d*. Up 1*s* 1*d*. The cheapest region for bacon, probably due to the concentration of bacon factories in the area. A lb of bacon was up to 4*d* cheaper than in London.

North and Scotland: £4 17*s* 7¾*d*. Up 1*s*. Fewer cut-price, self-service stores and supermarkets probably account for the North and Scotland having to pay most for the shopping basket. The North is now having to pay 5*d* a lb more for tomatoes than the South and 3*d* more for cheese.

A Guide to 'unfrozen' prices

	Average price now	% increase since Aug.
Eggs (1 doz standard)	4*s* 7¾*d*	+28.0
Tomatoes (1 lb)	2*s* 4½*d*	+21.7
Dessert apples (1 lb)	1*s* 6¼*d*	+9.0
Cheese (½ lb)	1*s* 10½*d*	+2.4
Bacon (1 lb)	5*s* 1¾*d*	+2.3
Fresh fish (1 lb)	4*s* 0*d*	+2.2
Coal (½ ton)	£7 1*s* 1¾*d*	+7.0
Best bitter (1 pint)	2*s* 1¾*d*	+1.6

CROPS and MACHINERY

Tax muddle stops farm machine sales

John Winter 10.2.66

The muddle over investment allowances for farmers has caused a pile-up of new machinery at factories. Farmers refuse to buy until the position is made clear. The Government's handling of the £200 million-a-year farm machinery industry was described yesterday as a "direct insult".

Mr Lance Parker, a director of Massey

Ferguson, asked the National Power Farming Conference at Brighton to demand that the Prime Minister make an announcement. He said that before January 17, when the old system of allowances ended, a farmer could claim an initial tax allowance of 10 per cent, plus an investment allowance of 30 per cent. This meant that on a tractor costing £1,000 he could offset £370 against tax.

When new allowances for other industries were introduced the Government said that special arrangements would be made for farmers, but did not say they would be retrospective.

Mr Parker said: "The farmer planning his machinery buying does not know what his costs will be. This has brought buying to a halt and upset production planning by machinery makers. Our business is very dependent on seasonal demand. If you miss this demand one year the machinery is on your hands until the following year, and in our industry the word 'loss' is spelled RUST."

An informal meeting of farmers and machinery makers later decided to make representations to the Ministry today, through Sir Harold Woolley, NFU president, and Mr Gilbert Hunt, president of the Agricultural Engineers' Association.

The Ministry of Agriculture said: "The Government fully realises the importance of announcing details of new investment incentives for agriculture at the earliest possible moment."

All hands to the plough

John Winter 7.3.66

Many farmers plan to work dawn to dusk in their fields this week to try to make up for one of the worst starts to a year they have ever had. Incessant rain has held up cultivation for three months in many areas. Cereal sowing and potato planting are three to four weeks behind, and thousands of lambs have been drowned or have died soon after birth in waterlogged fields.

Although these losses cannot be replaced, a mainly dry spell in March would enable arable farmers to catch up, though cropping programmes will have to be changed. In December winter wheat, at 1,500,000 acres, was nearly 500,000 acres behind the previous year, and it is now too late to sow it. This means farmers will have to substitute spring varieties, or switch their fields to barley.

Hardly any potatoes have been planted, even in the earliest growing areas. Last year at this date 3,000 acres were in the ground in Cornwall and 9,000 acres in Pembrokeshire. Mr Melville Scoble, Cornish member of the Potato Marketing Board, said: "The next two weeks are vital. With good planting conditions and warm growing weather the crop could still be ready for lifting at the normal time. But with planting going on everywhere at once, instead of being staggered, too many crops will be ready at the same time."

In the East Anglia cereal belt many farmers have still to complete their ploughing, and only a small amount of spring corn has been drilled.

Experts take the headache out of weeding

John Winter 31.3.66

Twenty weed "detectives" went round a field of winter wheat on their hands and knees at Bagshot, Surrey, yesterday. They found mayweed growing at the rate of more than four million plants to the acre, knot grass at more than three million, 12 other weeds in significant numbers and traces of 20 others.

This season thousands of farmers have handed over their weed headaches to the experts. They also save money, say weed specialists of May and Baker, the Dagenham chemical manufacturers who operate this "leave-it-to-us" service.

Mr Laurie Taylor, head of the firm's agricultural sales, said: "Farmers have often bought weed killers in advance only to find, when the weeds appear, that they have got the wrong chemicals for the job. This service ensures that they have the right chemical, with instructions on how to apply it at the right time."

Each farmer is visited by a specialist who examines every field, and the success of the scheme depends largely on his ability to recognise weeds in the early seedling stage and make correct recommendations for their control. Several hundred thousand acres of cereals, root crops, peas and even grassland will have been inspected before the end of April, when weed spraying will be at its peak.

Push-button magic drives tractor of tomorrow

Peter Bullen 20.5.66

With one finger I drove the tractor of tomorrow around the ring of the All-England jumping course at Hickstead, Sussex, yesterday.

By operating a push-button electronic control box no bigger than a typewriter I put the tractor through its paces: starting, accelerating, changing gear, reversing, stopping, sounding its horn. No one was in the driving seat. The tractor was activated by radio waves from the control box with its 28 buttons.

The tractor, an ordinary production-line Ford 3000, has been adapted to respond automatically to the signals. The magic box makes tractor driving as easy as switching on the television and selecting a channel, and it can do any operating a human driver can do. The control box has a built-in "fail-safe" system which stops the tractor when anything goes wrong or it goes out of range, which is about 300 yards. It is the most advanced remote controlled vehicle of its kind in Britain and it has taken ten years of research to produce such a comprehensive control system.

But farmers are not being sought as customers. The main commercial outlet at first is likely to be for jobs which are too dangerous for human drivers. Suggested uses are for atomic energy stations, civil defence work, fire-fighting, oil and chemical works, and military uses like mine and bomb disposal. Government observers yesterday were impressed when they saw the tractor driven blind through a tunnel, carrying imaginary radioactive material with its front loader. It also carried real explosive which was used to blow out a simulated oil-rig fire.

Full-scale adaption of the tractor to remote control costs about £3,000 and takes six to nine months but both price and time could be reduced if orders warranted it. Simpler and cheaper systems which would appeal more to farmers could easily be produced.

Farmers make their own rain

Peter Bullen 24.5.66

More farmers are using artificial rain to boost their crops or to protect them against severe frosts. The number of farms using irrigation went up by more than 200 between 1963 and last year. The acreage watered rose from 211,000 to 266,000.

One of the most significant developments in the past two years has been the rapid growth of farmers' own reservoirs. There were reservoirs on 439 farms in 1963, which held 546 million gallons of water. Last year the figure almost doubled – 834 containing 1,058 million gallons.

More farmers and market gardeners are using irrigation equipment to spray fruit trees and plants as a protection against late frosts.

Rain hits biggest corn crop

John Winter 4.8.66

Persistent rain is holding up the biggest corn harvest ever grown on British farms. In forward growing areas of the South, where cutting of early crops started last month, harvesting has stopped. Heavy rain in the past few days has beaten down thousands of acres of standing corn, and lack of sunshine is delaying ripening.

The June farm census returns for England and Wales, due today, are expected to show that, while the wheat crop will be more than 200,000 acres less than last year, barley will have jumped by about 600,000 acres. The bulk of this crop is spring sown, and bad weather then resulted in much of it going in late. In the good growing conditions of early summer the late crops made rapid headway, but the present poor weather is retarding them. Unless they get plenty of sun in the next week or two they will suffer in both yield and quality, and farmers will face a long, late harvest.

In Sussex barley already sold to a Chichester corn merchant has yielded well, with some crops up to two tons an acre, but yields of late-sown spring barley are expected to be much lower. A merchant in Hampshire forecast the heaviest harvest he has ever handled, with yields better than last year – provided the sun shines.

The National Farmers' Union in the South-West said: "Farmers are watching the weather anxiously."

Bumper crop on the way

John Winter 18.8.66

Combines moved into the Cotswold cornfields yesterday for the start of a harvest which could be the best for some years on this high, windswept land. Three days of warm sunshine have transformed crops which a week ago seemed a long way from harvesting condition. Storm damage, which often flattened huge areas of cereals in the region, is comparatively light and, given a continuance of good weather, the farmers look forward to yields of good weights and quality.

Mr Charles Whittem, who started to combine a 25-acre field of Rika barley near Stow-on-the-Wold, said: "I tested this grain on Sunday and it was quite soft. Today I could not bite through it. It is the first patch of 240 acres of barley and wheat which I have to combine, and if the weather holds I shall be able to go right ahead. I am starting 10 days earlier than last year. This field, which was then sown with wheat, got into such a mess that it was never harvested." He estimates the yield of his first field at about 30 cwt an acre, which is considered satisfactory for the area.

Over the West Midlands as a whole harvest prospects are fairly good, with yields up to average or a little above, and quality better than last year's rain-damaged grain. Owing to bad weather in the autumn, followed by a cold wet spring, there is a tremendous swing from winter wheat to spring barley, much of which was sown very late. The resultant crops look well in the field, but yields may be disappointing.

 # LIVESTOCK and POULTRY

News from Chi-Chi honeymoon home worries Britain's farmers

Peter Bullen 3.3.66

No admiring Russian crowds will see London Zoo's giant panda Chi-Chi and her boyfriend An-An at Moscow Zoo. Chi-Chi, who leaves England later this month, will find the Moscow Zoo deserted, apart from keepers.

All visitors have been banned because a virulent form of foot-and-mouth disease has spread across Russia and zoo authorities fear that visitors from outside the city could take disease into the zoo. Many valuable cloven-hoof animals could be affected, though it would present no threat to the pandas.

London Zoo said yesterday: "Chi-Chi's trip is still on."

In Britain vets and farmers are watching anxiously as the disease once again spreads across Europe. Every few years the disease suddenly builds up to epidemic proportions similar to the waves of influenza that sweep across countries. In the past two or three months there has been ominous evidence that 1966 could become another peak year ... and the death of hundreds of thousands of cattle, sheep and pigs would have dire effects on Europe's meat supplies. It could lead to another year of high meat prices in Britain if Continental buyers come across in force to buy English cattle or pay high prices to attract meat from our traditional suppliers like the Argentine.

Recently there have been thousands of outbreaks of foot-and-mouth disease in Europe – more than for several years, but nowhere near as bad as the early 1950s when outbreaks ran into hundreds of thousands. One of the world's leading authorities on the disease, Dr J B Brooksby, director of the Animal Virus Research Institute at Pirbright, Surrey, said: "The position is serious. Just how serious we cannot tell until we see further figures on numbers of outbreaks."

There are three strains of the disease which are causing concern, he explained:

• *Continental "O"* type virus which caused hundreds of outbreaks in Germany after spreading from Hungary, Rumania, Austria and Switzerland.

• *Continental "C"*, which had been prevalent in pigs for some time in Holland, flared up in cattle in late 1965 causing about 1,000 outbreaks in December and 1,422 in January, and leaked into Belgium, where there were 145 outbreaks in January.

• *Near East "A"*. Dr Brooksby was deeply concerned about this type of virus, which is much more virulent than the other two. Last year it spread through Central Russia, and there is now a "massive concentration" in that area. It went through Greece, Turkey and Bulgaria and was contained at the borders with Europe only by a large vaccination programme in the same way as another virulent African virus was stopped in the same countries in 1962–63.

It is still smouldering in these countries and in Russia, with its long land boundary with Western Europe. Dr Brooksby said: "There is great concern about what may happen in Russia when winter is over and cattle leave their winter quarters to come into the open."

The threat of foot-and-mouth being introduced into Britain will grow now with the warmer months bringing foreign visitors, returning holidaymakers and migrating birds from the Continent. Already the Ministry of Agriculture has banned meat from Holland and Switzerland – foot-and-mouth virus can live in bones and bone marrow for months. Meat used in pig-swill, or an infected bone taken by a dog into a field could cause an outbreak.

Farmers have been warned to look for signs of the disease and to take care in handling and cooking pig-swill. The Ministry's own vets in South-East England are keeping special watch, too, especially at ports. And the Ministry advises farmers who have to go to Europe: "Keep away from livestock farms and markets."

Years of hard work to eradicate foot-and-mouth from Britain would be spoiled if the disease was introduced. For 3½ years now – apart from a solitary outbreak in Kent about a year ago – the country has been free from the disease. This is due to a ruthless but very effective policy of slaughtering all animals in the vicinity of an outbreak and compensating the owners at a cost to the Exchequer of anything up to £2.5 million in a bad year.

Britain has now done almost everything which can be done to prevent the disease coming in and once again the English Channel is playing its ancient role as the last barrier against invasion.

Super pigs – 8,000 a year for sale to breeders   John Winter 20.4.66

Britain's biggest manufacturer of feeding stuffs is planning to become the country's biggest pig breeder.

British Oil and Cake Mills, the £70 million-a-year Unilever feed firm, has started a farm in Yorkshire which in five years will be producing 8,000 super pigs a year. They will be hybrids, derived from three pure breeds which have been computer-selected to give a 25 per cent improvement on present-day pigs in rate of growth, food conversion, and the balance of lean and fat which customers prefer in pork and bacon. Most will be sold by the firm at commercial rates to pig farmers. The others will be fattened and sold for slaughter.

Mr David Bellis, BOCM pig expert, said yesterday: "The main reason we have entered pig breeding is to provide animals which are capable of making the best use of scientifically developed feeding rations. Feeding a poor pig with the new rations is like putting 100-octane petrol into a 1925 car. Our aim is to breed pigs as efficient as modern cars, and so boost the incomes of farmers and keep the industry prosperous."

The company had no intention of entering pig production except to provide improved stock.

Already more than 3,000 test pigs are being raised on 27-acre Blackwood Farm, near Selby, which three years ago was open fields. In the first stage, litters of two pure breeds, Landrace, noted for long, lean carcasses, and Wessex Saddleback, reputed to be the best mothering breed, are being tested. Records of weight and food consumption, and measurements of fat and lean made by ultrasonic "sounding" on the pigs' backs, are fed into a computer. This selects only the best 12½ per cent of females, and a smaller number of boars for breeding. The rest will be fattened and slaughtered.

The cream of the two breeds will be mated and their female offspring later mated to Large White boars.

Heavyweight Cassius loses on points
John Winter 20.5.66

Cassius Clay, a loud-mouthed, heavyweight, golden bull, was beaten on points and toppled from his champion's throne at the Devon County Show, Exeter, yesterday.

Cassius, a six-year-old weighing in at 24 cwt, is entered in the South Devon breed herd book as Edmeston Cracker. He won the breed championship at last year's Devon Show and he earned his nickname from his habit of bellowing all the time he is being led round the ring. His owner said: "Just to let everyone know how good he is."

He did it again yesterday, but it got him nowhere. The habit caught on and his successful rival, Wishworthy Flash 12th and three other competitors all opened their mouths and roared in unison.

Mr Jim Hosking, the judge, said: "It was a remarkable sight and sound." It was also one of the finest turnouts of senior bulls I have ever seen and there was very little to choose among the first three. "They all carried plenty of beef and were the sort of animals to breed fine beef calves. I thought Cassius (placed third) was not at his best."

Flash, six years old next month, also scales 24 cwt. He was shown by Mr John Warne, who farms with his father at Tregony, near Truro. They have a herd of 60 South Devon cows, plus young stock, and breed from them to produce both rich creamy milk and lean tender beef. Mr Warne said: "I suppose we should start calling our bull Henry Cooper."

The dethroned champion was bought after the last show season by Mr Bernard Thomas, of Tregerrick, Gorran, Cornwall, who has an 80-head herd on a 260-acre farm.

Government plans drive to stamp out cow disease
Peter Bullen 27.7.66

A scheme to eradicate brucellosis disease in cows is to be introduced by the Government as soon as possible, Mr Fred Peart, Minister of Agriculture, promised yesterday. The disease, also known as contagious abortion in cows, can lead to illness and sometimes death in humans who drink milk from affected animals before it has been pasteurised. Brucellosis causes heavy losses to dairy farmers and is a threat to our export trade.

Mr Peart, in a written Commons reply, said that the scheme would not only serve the interests of human and animal health, but would also improve agricultural productivity and enable us to maintain and increase exports of livestock. "The first essential is to build up a register of brucella-free herds on a voluntary basis to provide a reservoir of disease-free replacements. The second stage, which can be introduced only when the voluntary response is large enough, will consist of a plan of eradication, area by area, in which all animals reacting to diagnostic tests will be slaughtered, with payment for compensation."

Various aspects of the plan, to be introduced over a period of some years, have to be discussed with the NFU.

Baroness Summerskill, opening a debate in the Lords, said that raw milk potentially infected with brucellosis could still be sold from vending machines. Calling for action, she said: "It is a criminal confidence trick. A woman will go to a vending machine because she is told milk is good for her child and she is getting milk which might infect her child with brucellosis."

The debate on the need to eradicate the disease came a few hours after Mr Peart's announcement.

Charter to protect farm animals
John Winter 6.8.66

A Charter to protect chickens, pigs and cattle from overcrowding and suffering in factory farms is to be drawn up by the Government. Its standards will not be strictly enforceable, but farmers who ignore them may be prosecuted for causing unnecessary suffering.

Plans to stop or limit many methods which opponents of factory farming have branded as cruel were outlined to Parliament yesterday by Mr Fred Peart, Minister of Agriculture. Legislation will be introduced to give wide powers of control over the way animals are housed, fed and treated.

While the basic approach will be through a comprehensive code of practice, to which farmers will be expected to conform, some practices will be banned. These include feeding calves on a diet deficient in iron to produce "white" veal; keeping pigs and poultry in houses with inadequate light for inspection; and dock-

ing pigs' tails to stop other pigs from biting them.

The most far-reaching new power will be that official veterinary staff will have right of entry into any farm to inspect welfare standards.

The Government will set up a standing advisory committee on all aspects of animal welfare. It will include representatives of farmers and welfare organisations, as well as scientists and vets. The committee will work out codes for each kind of production, which will include:

• *Poultry*. Adequate space, feeding and drinking facilities for each bird. Removal of part of the beak to prevent birds pecking each other will be discouraged, and probably banned later.
• *Pigs* will have minimum floor space standards.
• *Calves and adult cattle* will also get adequate living room, and those kept tethered, yoked, or in stalls, sufficient freedom to groom their flanks.

A surprising omission from Mr Peart's plans is specific reference to battery cages, in which more than half of the 30 million laying birds in England and Wales spend their lives. Batteries are the main targets of criticism among anti-factory farming campaigners, and the Brambell Report recommended that no more than three birds should be kept in one cage.

Mrs Ruth Harrison, the Kensington housewife whose book, *Animal Machines*, led to the creation of the Brambell Committee by Mr Christopher Soames, former Tory Minister of Agriculture, said yesterday: "We are back at square one. Absolute standards should be laid down in legislation. We will fight this very hard. I am particularly disappointed that there is no recommendation on batteries."

Half of the farm vets may have cattle disease *John Winter 29.9.66*

Half the veterinary surgeons who treat Britain's farm animals may be suffering from a disease contracted from cattle. Brucellosis, which causes cows to abort, is known in human cases as undulant fever. Symptoms may range from sweating and lassitude – it is sometimes mistaken for influenza – to rheumatism and arthritis.

The extent of its incidence among the country's 5,000 vets will be revealed today in a report to the British Veterinary Association Congress in Brighton. The report, by two university researchers and two public health officers, shows that of more than 300 vets examined in a survey nearly 200 showed evidence of infection. More than 50 had symptoms suggesting chronic brucellosis. Mr Alasdair Steele-Bodger, Veterinary Association president, said earlier in the congress: "The report will make your hair stand on end." One of the compilers of the report, Dr William Kerr, of Belfast University, cannot attend the congress because he is himself suffering from chronic brucellosis and will be off work for four months.

Brucellosis has always been accepted by vets as an occupational hazard, and for years they have been pressing the Government to eradicate it by slaughtering infected cows. In July, Mr Fred Peart, Minister of Agriculture, announced a scheme to compile a register of disease-free replacement animals as a preliminary to eradication. But a start on stamping out the disease is not even in sight.

Britain bans 265 French blondes *John Winter 6.10.66*

The biggest consignment of foreign cattle into Britain – 30 bulls and 235 heifers of the big blonde Charollais beef breed – has been cancelled after a row between British and French Government vets.

The French refused to accept a new test for foot-and-mouth disease devised by the British as an added precaution following the Northumbria and Sussex outbreaks.

The animals, worth £150,000 in cost, transport and quarantine charges, were due here before Christmas. The cancellation could mean the end of five years' infusion of French blood into British beef. The ban is a blow to the young British Charollais Cattle Society, whose members aimed to add a new pure breed to the 20 we already have. While some of the bulls were destined for the artificial insemination service, most were ordered by breeders who already have a herd nucleus, and by 31 farmers new to Charollais breeding.

Society chairman Mr Anthony Harman, of Chesham, Buckinghamshire, said: "We have placed orders for the cattle and deposits have been paid, but not very much has been spent so far. I prefer the hold-up to come now, rather than after the cattle have gone into quarantine, and a ship chartered."

Quarantine regulations normally stipulate a month in a French station at Brest, during which a series of disease tests are made, and another month for further tests at the British port of arrival. Previous Charollais imports have been dogged with disease troubles. The imports started only after a two-year battle by British breed leaders to keep the French breed out.

Hereford herd brings record £230,700
John Winter 18.10.66

The world's most famous herd of Hereford beef cattle was sold yesterday for a record £230,700 and will be dispersed to breeders in three continents. Farmers from every beef producing country in the world flocked to the Herefordshire village of Marden to watch a sale of 136 white-faced aristocrats of the Vern herd owned by the late Captain Richard de Quincey, the master breeder, who died last December.

In a five-hour non-stop marathon, auctioneer Mr Bill Gallimore knocked down the animals at the rate of one every two minutes. Top price and record for the breed at auction was 15,000gns for a yearling bull, Vern Scorpio. He was bought by Major Michael Symonds, of Hereford, on behalf of a three-man syndicate. His partners are Mr Elwyn Jones, of Hay-on-Wye, and Mr George Wheelwright, of Green Gulch, California. The bull will remain in Britain for two years.

Mr Wheelwright also bought four cows, including 14-year-old Dragee Plum Vern 12th, the oldest animal in the sale, which he will present to Mrs Barbara Turner, an old friend of Captain de Quincey and supporter of the Hereford breed.

The other animals bought for South Africa, Canada and Argentina, brought total exports from the sale to nearly £50,000. Second highest price, 12,000gns was paid by Mr Leslie Lewis, of Dilwyn, Herefordshire, in partnership with Mr Arthur Mucklo, of Stourbridge, for a two-year-old bull, Vern Rooti.

Dr John Phillips who has bought the 450 acre Vern Farm for £131,000 paid 8,500gns for the senior bull and 1965 supreme champion at the Royal and other shows, Chadshunt Trooper, and his purchases of 15 other animals brought his total bids to £32,000. A price of 4,500gns paid for eight-month-old Vern Sirdar was another record for Hereford calves. Altogether the 136 lots made an average of nearly £1,700.

Among the crowd of 5,000 who flocked to the picturesque farm among the autumn tinted trees of the Lugg Valley were breeders and cattlemen from all parts of Britain. Americans in Stetsons jostled with Argentinians, Australians, South Africans, Rhodesians and Canadians for room on 2,500-seater stands rigged round the sale ring in a field behind the farmhouse.

Letter of the week – donkey guard
Miss Anne Lumsden, Forest Row, Sussex

Sheep-worrying by dogs can be prevented by keeping a donkey in the field.

We have a lot of trouble with dogs here by Ashdown Forest, but thanks to my donkeys I have been able to run sheep without a single case of worrying. I have watched a dog chased out of a field seven times in an afternoon. This dislike of dogs must be a natural instinct in donkeys.

Victory will be the death of Moreta
John Winter 5.12.66

A few white hairs tucked away invisibly in an otherwise all-black coat have condemned to death the favourite for the title of the best beef animal in the world. Pride Moreta of Thorn, aged 20 months, is tipped for the supreme championship of the Royal Smithfield Show, which opens its five-day run at Earls Court, London, today. Last week she was champion of her native Scottish National Show at Edinburgh, and survived to compete again. But if she repeats her success at Smithfield, nothing can save her. The decision of the Royal Smithfield Club to make auction and slaughter compulsory for all animals which get into the supreme championship list follows the last-minute reprieve of the supreme champion last year.

After the animal, a Hereford heifer, had actually been bought by a butcher for £2,000, her owner persuaded him to sell her back for breeding and she has since produced a calf. "I would not have considered entering Moreta under this rule if she had been suitable for breeding," said Mr Alan Grant, of Thorn, Alyth, Perthshire, who has twice before won the supreme award.

MP accuses Ministry of 'whitewash' over foot and mouth
Peter Bullen 16.12.66

The Ministry of Agriculture was accused yesterday of allowing the spread of foot-and-mouth disease through inefficiency and of using hush-hush methods to cover up blunders. The charge was made by Lord Lambton, Tory MP for Berwick, speaking at Acklington, Northumberland.

During July, August and September the stock on 32 farms in the north of the county were slaughtered, costing the country £800,000 in compensation. The outbreak "nearly ran like a flame through England," said Lord Lambton.

The Minister of Agriculture, Mr Fred Peart, has refused an inquiry – but Lord Lambton has made his own "painstaking investigation". This, he said, had convinced him there had been a deliberate attempt at "whitewash" by the Ministry.

"This in itself is serious; but I have also been conscious of deliberate attempts to hush the original errors up by putting pressure on those whose livelihood is dependent upon business provided for them by the Ministry. We have both concealment and inefficiency and now underhand pressure. All this points to the necessity for an inquiry."

The Ministry of Agriculture said last night that reports seen by the Minister "do not include any evidence to support the allegations attributed to Lord Lambton. If any such evidence can be produced, Mr Peart will of course consider it."

 # EMPLOYMENT

Farms death toll is down
Peter Bullen 3.1.66

The toll of farm deaths dropped in 1965. Ministry of Agriculture figures up to the last 48 hours of the old year showed 90 deaths through farm accidents. This compares with 106 the previous year and is less than two-thirds of the 1961 peak of 141 deaths.

Mr George Wilson, the Ministry's chief safety inspector, said: "Barring a crop of fatal accidents in the last few hours of 1965 the death toll looks like being the lowest we have ever had. Non-fatal accidents in the first nine months of the year showed a corresponding reduction.

"A substantial contribution has been made to farm safety by agricultural machinery manufacturers and by all sections of the industry, backed by recent safety legislation. There is every reason to think that the farming public are more safety conscious than ever before and this will help to continue the trend of fewer accidents."

Most disheartening aspect of the accident log was the persistence of two major groups of figures – the number of children under 15 who died and of tractor drivers crushed and killed by their machines overturning. Last year 16 children were killed, the same as in 1961, 1962 and 1963. In 1964 there were 18.

Peart acts over tractor safety

Peter Bullen 15.1.66

Within the next ten to 12 years all tractors will have to be fitted with safety cabs or frames, under proposed regulations which the Minister of Agriculture, Mr Fred Peart, published yesterday. He wants all new tractors to comply within two years of the new regulations becoming law, and existing tractors within ten years.

In the past nine years 300 drivers have been crushed to death under overturning tractors, about one-third of all fatal accidents on farms. Last year 36 of the 93 farm deaths were due to this reason. The Ministry said yesterday: "Investigations have shown that most, if not all, of the drivers' lives would have been saved if they had been protected by suitable safety frames or cabs."

Interested parties will have until April 14 this year to give their comments about the proposed regulations to the Ministry. The National Farmers' Union's reaction is likely to be support for any measure to stop tractors overturning in the first place by better tractor design and by education

and publicity for drivers. The NFU is not convinced that compulsory fitting of safety cabs is the best answer to the problem.

Letter of the week – tractor safety
L H Pratt, Nacton, near Ipswich 24.1.66

A very respected and experienced tractor driver in the village was killed last week when his tractor up-ended. This is the hazard you reported as being responsible for one-third of all fatal accidents on farms.

Although I am not personally connected with farming I am convinced the answer lies in front-wheel drive. There are a lot of snags, but I feel the idea is worth investigating.

Where have all the workers gone?

Peter Bullen 22.4.66

Harpendenbury Farm, Hertfordshire, is a ghost farm. Barley grows in large, orderly fields on most of its 350 acres tucked between the M1 and the little town of Redbourn. There are four stout cottages and a cluster of buildings around a rambling farmhouse of mellow red bricks. It looks like any other well-run, medium-sized arable farm.

However, for the past seven years not one regular farmworker has been employed or even lived there. The land has been ploughed, cultivated and fertilised. Each year the seed has been sown and the grain harvested – but every major operation has been carried out by workers supplied by an agricultural contractor. The work is done swiftly (one harvest was finished in 11 days) and the fields are left silent and deserted, apart from the regular rounds of the farmer, 38-year-old Mr Hugh Stovin.

The contract farming system he is pioneering is beginning to spread. A neighbouring farm and several others in Lincolnshire have switched to it. They are not small, one-man farms either. Thirteen years ago Harpendenbury Farm had nine full-time workers looking after cows, sheep, poultry and mixed crops. Now the livestock have gone and only barley is grown, apart from a few acres of herbage seed.

Mr Stovin has ripped out ten miles of hedges, put up a grain-drying unit and store and uses only liquid fertilisers to streamline the work. More farmers are being forced into similar moves by rising costs and falling prices. The Government is also encouraging them to modernise and enlarge holdings to make them more productive. Increasing mechanisation is helping, and so is the flight of 20,000 workers a year from the land.

Mr Stovin was laughed at when he advocated contract farming. He said: "When I got rid of all my workers in 1959 every farmer I met said I was mad. In the past few years many big farmers have told me they wished they had had the courage to do the same."

He pays the contractor about £3,000 a year but estimates this is £1,000 a year less than it would cost to employ two full-time farmworkers. It was not merely the savings in wages, he said. His paperwork was simplified without wages and tax problems, he did not have to have £10,000–£15,000 tied up in machinery which was little used, and his cottages were no longer a liability. Instead of costing him about 30s each a week for rates and maintenance when workers lived in them rent-free, he was now making about £500 a year by renting them to people who worked in nearby towns.

Contract workers who did the same job every day became much more proficient than regular farmworkers faced with several jobs. "A regular ploughman on my farm would do perhaps 300 acres a year, whereas the contractor's man probably does something like 2,000."

He hopes eventually to have even more time to spend with his wife and three-year-old son. At present he is toying with the idea of cutting out all ploughing and using paraquat, the "chemical plough". "I could even become a part-time farmer," said Mr Stovin, who practised as a barrister for two years before taking up farming. I am beginning to get things in order and perhaps by the time I am 40, I'll have enough time to spare to find other employment."

Extra skills will give extra pay in NFU plan   *Peter Bullen 11.6.66*

A plan to pay farm workers extra money for special skills and responsibilities will be put to the National Farmers' Union wages committee next month. If approved by the committee and later by the NFU council it will be put to the Agricultural Wages Board for discussion with workers' representatives.

Since 1961 the National Union of Agricultural Workers has been pressing the board to bring in a wages structure and has suggested a system of special payments for the different jobs a farm worker has to do.

The NFU wants the emphasis to be put on the man, not the job, to avoid "who does what" arguments and has suggested grading workers into four groups. *Group one* would be for unskilled entrants receiving the basic wage; *Group two* would receive a bonus for special skills; *Group three* would get even more for jobs with responsibilities like looking after a herd of pigs or cattle; and *Group four*, the best paid, would be for men with farm manager's responsibilities for whole farms.

Day-off pay claim thrown out

Peter Bullen 29.6.66

Farm workers' demands for extra overtime payments for working on days off were rejected by the Agricultural Wages Board yesterday. The workers wanted double time instead of the present time-and-a-half for working on Sundays, Bank Holidays and their weekly short day.

The farmers' representatives on the board said that to pay 9s 8d instead of the present 7s 2d an hour overtime to 400,000 farm workers would cost an extra £4 million a year, which they could not afford. They also said that the workers had a 6.3 per cent wage rise and an hour cut off the working week in January.

Lord Collison, general secretary of the National Union of Agricultural Workers, who led the workers' representatives, said afterwards: "Naturally, I am disappointed. We thought the farm worker had a good case because he has to work on Sundays to look after stock. He can't get away from the farm. This dedication ought to have been appreciated."

Lord Collison said that the workers' representatives would meet again on Monday to consider a claim for a cut in the working week from 44 to 40 hours and an increase in the minimum wage from £10 10s to £14, which was demanded by the union's conference last month.

Land workers seek big pay rise

John Winter 1.9.66

Farmworkers' leaders yesterday put in the biggest pay claim of any union since the wage freeze started. The claim for a "substantial rise" for 430,000 men and women, is based on the demand at the National Union of Agricultural Workers' conference that the present rate for full-time men should be raised from £10 10s to £14 a week.

Union officials who put the claim to the Agricultural Wages Board accept that under the freeze no increase could be paid before the end of the year. But the board could make an award payable from next January when the freeze ends. Claims first presented in July in the past two years have brought increases payable from the following January.

Lord Collison, NUAW general secretary, said: "We support the wage freeze, and our application is in full accord with Government policy."

The union's two main points are the widening gap between wages of farm-workers and those in other industries, and the flight of workers from the land into more highly paid jobs in towns. Hourly earnings are only two-thirds of the average in other industries – 5s 6½d against 8s 4d and weekly earnings are £13 7s 8d as against £20 5s.

Farmers' leaders will reply to the claim on October 14.

Training farm boys may cost millions

Peter Bullen 29.9.66

The first step towards setting up a complete, modern training system for Britain's 400,000 farm workers was taken yesterday. It happened at the first meeting of the Government-initiated Agricultural Training Board. The board – on which there are 30 farmers, workers and educationists – hopes to improve training standards in agriculture, horticulture and forestry and to ensure efficient facilities all over the country.

The cost may run into millions of pounds a year, raised by a levy on the 150,000 employers. Chairman Mr Basil Neame said they had no idea how much the board would cost to run or what a levy would bring in. Similar boards had raised payroll levies of between ½ to 2½ per cent of total wage bills. Agriculture's wage bill is £330–£340 million a year.

The first job would be to appoint a director, then training officers, probably one for each county. Next, the 150,000 employers would be listed and persuaded to support the board.

The board would run courses, fix standards and tests and help people find training facilities. A major part of its expenditure would go on grants to farmers providing approved training.

Farm accident deaths rising

Peter Bullen 21.11.66

Fatal accidents on farms have increased this year. With figures for almost two months still to be returned, the toll stands at 100, seven more than in the whole of 1965.

The Ministry of Agriculture's chief safety inspector, Mr George Wilson, appealed yesterday for farmers and workers to take even more care than usual. He said: "At this time of the year, particularly, more care should be taken both indoors and out, because the light is not so good and weather conditions can lead to treacherous working surfaces."

Overturning tractors were again the biggest killer. So far this year 34 drivers have died. Every year the Ministry is concerned about the constant number of about 16 children who are killed on farms. This figure was reached in the first ten months of 1966.

There were fewer non-fatal accidents in the first six months of the year. And there has been a considerable reduction in the number of accidents caused by bulls, probably due to the greater use of artificial insemination which has led to a cut in the number of bulls on farms.

Farm workers settle for a 6s rise

John Winter 11.12.66

A wage rise of 6s a week – just under 3 per cent – was awarded to farm workers last night. It is the first major increase negotiated for an entire industry since the freeze began. The rise, payable from February 6, was decided by the Agricultural Wages Board after a seven-hour discussion, in which the workers' representatives at first claimed a 10s increase. It is subject, under the Prices and Incomes Act, to approval by Mr Fred Peart, Minister of Agriculture.

The 6s award is for full-time workers. But, with proportional increases for women, juniors and part-timers, 400,000 farm workers in England and Wales will benefit. The increase, which will raise the minimum wage for a 44-hour week to £10 16s, was a disappointment to the workers' leaders. They claimed that under the Government's recent White Paper the farm men qualified for an early increase during the six months of severe restraint beginning in January.

Lord Collison, general secretary of the National Union of Agricultural Workers, said: "Taking into account all circumstances, including the Government's economic policy, we considered 10s appropriate. We think we were justified because of the farm workers' record of raising productivity by 6 per cent a year – double the national average."

The increase is the smallest farm workers have received for nearly five years. Lord Collison, who thinks it will be approved by the Minister, said he was sure the NUAW would make a new claim as soon as the national economic position improved.

 PEOPLE in the NEWS

Shall I stay? asks Woolley

John Winter 7.1.66

Sir Harold Woolley, National Farmers' Union president, has asked all his 144 council members and 59 county branch chairmen if they want him to change his mind about retiring from the leadership in April.

As reported in Farm Mail on Tuesday, Sir Harold, 60, who is recovering from a mild slipped disc at his farm at Hatton Heath, near Chester, has been inundated with messages from leading colleagues throughout England and Wales. Many emphasised that his departure at this time would be a disservice to the industry.

Some branches have suggested that he should stay for one more year, his seventh, to allow more time for the union to consider possible successors, or even a change in the leadership pattern. The top job has become so complex and time-consuming, it is said, that the union should look for a highly paid, professional "Beeching" to fill it.

Sir Harold, who expects to be back at work, fully recovered, next week, said: "I have been deeply impressed by the messages I have received." In his letter he carefully avoids saying whether he would be willing to stand for re-election, though he indicates that his great enthusiasm to serve farmers and the industry has not weakened.

Woolley to stand again as president

Peter Bullen 18.1.66

Sir Harold Woolley, 60, yesterday reversed his decision to retire as National Farmers' Union president. He told a council meeting of more than 100 delegates from every county in England and Wales: "Perhaps I was a little hasty when I said at the last council meeting in December that I would retire."

Sir Harold told me later: "Perhaps I had been over-sensitive as to the extent of the support for my leadership and certainly the leader of this organisation needs a great volume of support." Sir Harold said he had had "an overwhelming" response from members to reconsider. If he left now it could harm the union which was going through a stormy period.

Now Sir Harold will have to wait until April 7 to see if the support he has had in messages from council members and county branch chairmen measures up to the 85 per cent vote he needs to be re-elected.

Sir Harold fails to get backing for election

John Winter 24.1.66

It looks as if the National Farmers' Union will elect a new leader in April after all. County branch chairmen and national council delegates have not offered enough support to Sir Harold Woolley to justify his running for a seventh year as president.

John Winter MBE, agricultural correspondent of the Daily Mail from 1958 to 1979.

A kind-hearted family man, John Winter's capacity for long hours of work was apparently fuelled by small cheroots and the pipe he smoked most of the day.

Two giants of post-war agriculture photographed in 1950. The Rt Hon Tom Willia
(left), Minister of Agriculture 1945-51, presents the Thomas Baxter Trophy to not
livestock scientist Dr John Hammond.

Christopher Soames (right), Minister of
Agriculture 1960-64, photographed in 1960.

Fred Peart (right), photographed
1967, was Minister of Agriculture
1964-68.

...dwyn Hughes on the left, Minister of Agriculture 1968-70, photographed in 1968
...h the Daily Mail's Peter Bullen. In the background are some of the Guinness
...d of Guernsey cattle.

The two Presidents who led the NFU during the 1960s.
On the left, Harold Woolley; right Gwilym ('Bill') Williams.

Harold Collison, Secretary of the NUAW, addressing delegates on the second day of the 95th Trades Union Congress in Brighton, 1963. He was speaking about social insurance and industrial welfare.

This photograph by Chris Barh 1967, was captioned 'The Jones

'Potato rebel' Jack Merricks, right, inspects a crop of King Edwards in October 19

mpanied John Winter's account of the foot and mouth disaster on November 17,
deserted farm with Nell, the collie, the only animal left alive'.

e tragic look on the face of Sylvia Jones (top picture) touched the heart of
mer Walter Bromfield who offered them free a bull calf to re-start their herd.
s picture is of his herdsman's son, Robert Bartlett, aged two, feeding one of the
ch from which the gift calf came.

A constant theme of the 1960s was the flight from the land by farmworkers, due either to low wages compared with industry, or the reduced need for labour follow mechanisation. The labour-intensive sugar-beet handling in 1963 (top photograph) contrasts with the mechanised lifting of the Standen harvester in 1969.

p) A traditional view of the 1960s: Proctor barley being drilled on previously
ughed and cultivated land in March 1964. (Bottom) In November 1966 this
totype direct draught machine was drilling barley into previously sprayed
bble without any ploughing or other cultivations. Direct drilling had arrived,
nks to chemical weed control.

Farm Mail frequently returned to the weather and its influence on successive harvests. Here wheat is being combined in September 1965 during a fine spell.

The trend to mechanisation accelerated through the 1960s. In March 1967 John Winter noted the launch of David Brown's largest tractor to date, the 67bhp 1200 Selectamatic weighing 2½ tons. Behind it is a David Brown 3-furrow reversible plough.

This is in spite of evidence sent to all the 182,000 members a fortnight ago that Sir Harold and his colleagues have prepared plans to harass the Government if it imposes another unpopular price review.

Members were asked to say if they would support such measures as reducing supplies of milk and other produce, cutting expenditure on farm supplies and refusing to co-operate with the Ministry of Agriculture.

Sir Harold is expected to announce his decision to the council in London on Thursday. For most of the 144 members it will be the first meeting with their leader since his shock announcement at the December meeting that he did not intend to seek re-election.

After urgent pleas from many delegates Sir Harold, 60, wrote to all chairmen and delegates asking for their advice on whether he should stand. Many have not yet replied. Of those who have, a majority are thought to have advised him to stand again, but the majority is not big enough. Under union rules, a president seeking re-election must be backed by 85 per cent of the council.

Although only a minority of counties have revealed their hand, those which have shown politely that they would prefer Sir Harold not to stand again hold more than 15 per cent of the council seats.

Devon, biggest branch in the union with five council seats, has not decided whom to support, but has made it clear it does not want Sir Harold.

Confirmation on Thursday of his decision to retire will start intensive lobbying to sound out support for potential successors. Clear favourite is Mr G T (Bill) Williams, deputy president.

Rebel farmer joins wool board

John Winter 26.3.66

Mr Jack Merricks, the rebel farmer, is now a member of two of the five agricultural marketing boards against whose policies he has fought, even to the point of going to jail. Wool Marketing Board election results announced yesterday show that he has won the seat for the southern region, defeating the former holder, Mr R L Johnston, of Minehead, Somerset.

Mr Merricks, 55, who keeps 2,500 Kent sheep on farms totalling over 2,000 acres in Kent and Sussex, is already a member of the Potato Board. He defeated the vice-chairman, Mr Halbert Renwick, last October.

Mr Merricks said yesterday: "Sheep farmers now seem to realise that the Wool Board, with the help of the Ministry of Agriculture and the National Farmers' Union hierarchy, has been fleecing them long enough. Wool producers were shorn of another 2d a lb on their wool in last week's price review, which was accepted by the NFU."

He said that as a board member, he would defend sheep farmers' interests from a practical and independent standpoint. He would try to win back the rights of wool producers' co-operatives, which, he said, were now reduced to the role of collecting agents for the board. He once refused to sell his own wool to the board, which is a monopoly buyer, and held back his entire clip for a year.

Woolley loses top NFU post

John Winter 9.4.66

The long wrangle over the National Farmers' Union leadership has ended in the defeat of Sir Harold Woolley and the election of Mr G T (Bill) Williams as president. At the union council meeting

Sir Harold failed in two secret ballots to win the 85 per cent vote required by rule for re-election. In a straight fight after his elimination, Mr Williams, formerly his deputy, defeated Mr Henry Plumb, vice-president.

After his defeat Sir Harold listened to a string of glowing tributes from his colleagues.

The change at the top, only the second in 21 years, is the culmination of criticism from a number of county branches at some aspects of Sir Harold's leadership. The leadership switch will be followed by changes in policy, and perhaps in the union's structure.

Sir Harold, who was co-opted immediately after his defeat on to the council, said he would continue to work for the NFU. But he hoped to give much more time to his home, farm, and family at Hatton Heath, Cheshire.

The new vice-president is Mr Tom Cowen, 48, whose 330 acres at Thursby, near Carlisle, have been farmed by his family for 260 years.

Tough job, says new leader

John Winter 9.4.66

The new NFU president, Mr Gwilym Tecwyn Williams, has been on the NFU council since 1948 and began his first period as vice-president in 1953. He was born in Wales and speaks Welsh, but has been on his 450-acre farm at Longford Grange, Newport, Shropshire, for nearly 30 years.

For four months, while the leadership arguments dragged on, Mr Williams has worked daily with Sir Harold Woolley. He said yesterday: "I am very anxious to have a go, but I realise this is a tough job and there are no easy solutions to some of the problems which face the union.

Among his first priorities will be:

1. To decide how to meet pressure for more rapid advancement for bright young men.
2. Breakaway groups of dissidents, which are sniping at the NFU and damaging its image with rank and file farmers.
3. Improved communication between NFU headquarters and the men on the farms.
4. To be ready to consider any proposals to change Government policy on agriculture. ("Modification may be needed, but fundamentally the support system we have had for 17 years is sound.")

GENERAL

Country parks for town dwellers

Peter Bullen 1.3.66

The Government announced its plans yesterday to tidy up the countryside and open it up to more town dwellers.

Derelict cars and rubbish dumps will be banned. A network of new country parks close to towns and cities, picnic parks, rural walks and camping sites is to be created. Reservoirs, disused gravel pits and waterways are to be opened for boating, watersports and nature study. A Countryside Commission, formed by reconstituting the present National Parks Commission, will be responsible for promoting the countryside's leisure facilities.

Mr Fred Willey, Minister of Land and

Natural Resources, presenting the Government's White Paper *Leisure in the Countryside* yesterday, said he wanted the new commission to be a powerhouse, a dynamic body to stimulate countryside planning. The White Paper says the Exchequer will make grants of 75 per cent for work done by local authorities to promote enjoyment of the countryside and preserve its beauty.

Mr Willey said that at present £250,000 a year was spent on countryside amenities. He envisaged raising this to £2 million in a few years when the problems could be worked out and the money spent sensibly. Two focal points of the plan, he said, were water, which was relaxing, and trees, one of the glories of the English countryside, which they intended to preserve.

The movement away from towns in leisure time is growing. By the end of the century there will be 19 million more people in Britain and by 1980 today's nine million cars are expected to have increased to 26 million.

Many beauty spots and coastal resorts are too far from towns, driving there is slow and frustrating and too many other people have the same idea. Other areas might do just as well, but at present there is no reason to go to them, says the White Paper.

The Government hopes the new parks set up by local authorities will overcome these difficulties. They will be established throughout the countryside and along the coast. Where parks are not justified, small picnic places, "something better than a lay-by," with an acre or so of land and parking space, will be provided.

Special powers will be taken to regulate traffic in the countryside where it might endanger natural beauty or impair people's enjoyment of the scenery.

Farmers told they may have to take second place   *John Winter 18.4.66*

Farmers are becoming inured to the idea of rapid change in the countryside, which is their workshop. But they will need strong nerves to assimilate the revolution which is contemplated in a book* published today. They will be shocked even more to learn that far from being the most important in the countryside their needs may be subordinated to the urban millions who will need new and expanding playgrounds. These are the very people who depend for more than half of their food on the farmers whose domains they will invade.

But even practical considerations of survival do not seriously disturb Mr Garth Christian, an authority on countryside conservation, who believes that everything will work out right so long as children are taught the right approach to Nature at school, and all planners of the countryside are skilled ecologists. "Those concerned in any way with problems of land use need to know that recreation and wild-life conservation can create economic values, just as much as farming and forestry, though if man is to enjoy a pleasant environment economic factors cannot in every case be the prime consideration."

We shall have difficulty in reconciling this philosophy with the agricultural section of the National Plan. It is difficult to imagine the writer of a 200-page book on the future of the countryside making only one reference, and that somewhat

* *Tomorrow's Countryside*, by Garth Christian (John Murray, 35s).

derogatory, to the National Farmers' Union, but that is Mr Christian's achievement. On the other hand he mentions the Nature Conservancy more than 30 times.

His book has a foreword by Prince Philip, whose concern about the wise planning of our dwindling assets is well known, and brought the issues into welcome prominence in last year's Countryside in 1970 conference. Prince Philip writes: "We cannot afford to let things happen by chance or accident, only to be regretted later."

The book sets out clearly many facets of this complex problem, with some fascinating suggestions on how they could be met. It has admirable chapters on the exciting possibilities for our 1,500,000 acres of common land; the urgent need for protecting the shrinking stretches of unspoiled coastline; the compromise which must be made between demands for functional land-consuming motorways and the need to preserve unspoiled our winding country lanes; imaginative uses for abandoned railway lines, and the possibility of turning unattractive gravel pits and reservoirs into water sport centres.

Farm land goes up to £259 an acre   *Peter Bullen 16.9.66*

Farm land prices have started to go up again after the fall during the second half of 1965.

A report by Mr G H Peters, of the Agricultural Economic Research Institute, Oxford, shows that the average price for farms of all sizes rose by 13.7 per cent between January and June to reach £259 an acre. But this is only 3.6 per cent higher than the £250 average price reached in the first half of 1965.

The figures are based on records of 237 farm sales in the first half of 1966 compared with 234 and 258 sales in the two previous six-month periods.

There was a big jump of 15.3 per cent in the average price of farms in the 5–300 acres group due mainly to a rapid rise in the values of smaller properties between June 1965 and June 1966. In fact, farms of 5–50 acres reached £400 an acre compared with £318 and £351 for the previous two six-month periods.

Other prices: 50–100 acres £262 (£258, £233); 100–150 acres £238 (£239, £228); 150–300 £234 (£183, £233).

The only group where there has been no substantial recovery in values is for the over 300-acre farms. At £235 an acre they are still below the June 1965 figure although better than the £213 of the July–December period of 1965.

Big farm show killed by the squeeze   *John Winter 7.10.66*

The 105-year-old Royal Counties Show, one of Britain's major summer farm exhibitions, is dead – crippled by falling support and finally killed by the credit squeeze. Its 107-acre permanent ground at Kingsclere, Hampshire, bought for £31,000 two years ago, will be put up for sale.

Officials were authorised yesterday to negotiate with leaders of the Tunbridge Wells and Sussex shows to start a new South of England agricultural show at Ardingly, Sussex, next summer.

The winding-up decision came yesterday at a meeting of 60 of the Royal Counties Society's 2,700 members at Winchester. It followed a painful inquest on the failure of the last-ditch effort of the show – which used to move about its six counties of Surrey, Sussex, Berkshire, Hampshire, Dorset and Wiltshire –

to cut its losses and recover its fortunes on a permanent ground.

Mr Raymond Stovold, chairman, who also directs the booming Whit-Monday Surrey County Show, gave members a catalogue of disappointments: dwindling support by trade exhibitors, failure of a £3,500 drive for new members which brought in only 400 instead of the expected 3,000 and, finally, the impossibility of launching a drive for £100,000 capital in the current national economic plight.

Mr James Harris, treasurer, said that this year's show on three wet days in June lost £12,500, and the surplus of assets over liabilities shrank from £25,000 in 1964 to £17,000 in 1965, until today liabilities of £56,000 were just about covered by assets, which include the Kingsclere ground, now valued at nearly £38,000 in the books.

Mr Stovold said: "The plain fact is that we just have not got enough money, without a capital drive, to carry on another year. We have a very substantial offer for the ground. We have an offer from the Tunbridge Wells and Sussex societies to join them in forming a new South of England society on terms to be worked out. If we want to keep our good name rather than wind up and finish altogether this would be a sensible move."

The council's advice to abandon its own show met spirited opposition from a minority who wanted to fight on independently. Mr Richmere Worgan, of Basingstoke, accused the council of a "panic decision". And Miss M E K Pennington, of Alresford, said: "Once we leave our own permanent ground we are finished. I do not know how you can think of letting the Royal Counties go like this."

Mr Harry Mills, council member, retorted: "If you don't like it, what are you going to do? Do you want to gamble on next year?"

The meeting finally decided by 29 votes to 11 to abandon its own show, and then, without dissent, authorised its leaders to start negotiating today with the two other societies for a joint show next June.

1967

The appalling outbreak of foot-and-mouth disease dominated the news during the last three months of the year and it was to continue until the middle of 1968. It overshadowed the introduction of the Brucellosis Eradication Scheme. The establishment of the Agricultural Training Board and its despised levy gained most coverage on the employment front. Arable news identified "sustainability" for the first time and within livestock, embryo transplants and the birth of the first "home-bred" Charollais calf were both featured. The year also saw the death of Tom Williams, considered by many as the best ever Minister of Agriculture.

 ## POLITICS and FOOD

Peart is blamed for farm decline
John Winter 16.1.67

Mr Fred Peart, Minister of Agriculture, will be accused of starting a general run-down in British agriculture when farm leaders from all parts of England and Wales gather at the National Farmers' Union annual conference starting in London next Monday.

Resolutions say his policies have created uncertainty and falling profits for livestock farmers and loss of confidence in Government plans for agricultural expansion. They say the decline can be halted only by healthier profits from which farmers can invest for the future.

A Shropshire resolution, criticising the Minister for his handling of the collapse in beef cattle prices last autumn, says he shows "persistent inability to anticipate trends, or to take effective remedial action in time." It warns him that the spectacular rise in farming productivity will tend soon to turn into a recession, due to inadequate cash, credit and confidence.

Only 21 resolutions are listed, compared with 84 when the annual meeting was last held as early as January two years ago. On the third day the meeting will discuss two burning topics for farmers – British agriculture and the Common Market, introduced by Mr Bill Williams, president, and the home farmers' share of the home food market, which will be opened by Professor John Ashton, head of the agricultural economics department of Newcastle University.

Letter of the week – Roll on the Common Market

Robert Ashcroft, Rufford, nr Ormskirk, Lancs, 16.1.67

On January 10 last year I received £154 16s 3d for two grade 1 steers at my local market. This week at the same market for two steers of identical weight and grade, I got £124 4s 6d. I expect these figures will no doubt please the Prices and Incomes Board, but they certainly don't please me. Roll on the Common Market – we have nothing to lose.

How Europe would put up your food bills

Julian Holland and Peter Bullen 3.2.67

Now that Mr Wilson seems determined to take Britain into Europe, the time has come to look at the cost. What is going to happen to the personal budgets of Britain's 17 million housewives if we join the Common Market?

Of one thing there can be no doubt; whether or not our farmers get the alterations they want in the Six's farm rules – and this is the major bone sticking in the throat of the EEC negotiators – food will cost more. The National Farmers' Union estimates a family of four will have to find on average 25s a week more for food.

Take meat. Mr Peart, the Minister of Agriculture, has given these examples of present price differences: a joint of beef costing 7s 6d a lb in Britain costs the Germans 8s, the French 9s 7d, the Italians 9s 10d, the Dutch 10s 8d and the Belgians 11s 3d. Similarly, a piece of pork fetching 5s 6d in Britain also costs the French 5s 6d, but the Dutch pay 5s 11d, the Germans 6s 9d, the Belgians 7s 4d and the Italians 8s 6d. This will be the result of switching from our system of subsidising farmers with a guaranteed minimum price for their products to the Common Market system, under which consumers pay the full market price for food to provide farmers with an economic return for their work.

The German Minister of Agriculture, Herman Hocherl, says vital items in the British housewife's budget would cost 25 per cent more.

Mr Wilson has said that the increase in food prices will be from 10 to 14 per cent, though it is hoped the increases can be introduced gradually.

But if we were to go into the market tomorrow, just what would the housewife find when she went shopping?

The baker would charge her 2d to 3d more for a large loaf. Cakes would be dearer and flour would be up by 1d or 2d a lb. In the grocer's she would find most items up by at least 6d in the £. Sugar alone would cost 2d or 3d more for a 2lb bag. Butter would have shot up from about 3s 6d to 7s or 7s 6d a lb, and most cheeses would cost several pennies a lb more. Eggs would be 6d a dozen dearer, and bacon up 25 per cent. Prime back rashers now costing 6s would cost about 7s 6d.

Higher grain prices would be responsible, as they would for dearer beef and pork, for grain is the biggest item in the feed costs of livestock producers.

On the other hand, fruit – particularly from the Mediterranean – would be cheaper. Also some vegetables (like potatoes) would be down by about 1d a lb.

At the dairy, milk would be ½d a pint cheaper ... but a carton of cream now selling at 2s 3d would be 1s to 1s 3d dearer.

There would also be a saving of nearly £200 million a year on farmers' subsidies – though whether any of us would ever see this is another matter.

But, of course, dearer food is only one side of the picture. Against it must be set the likelihood of vastly increased prosperity if we join Europe; economic standstill and decline if we stay out.

Look, for example, at wages in Europe over the past eight years, the first eight years of the Common Market. Average hourly wage rates in the manufacturing industries have increased as follows:

Italy 87%
Germany 81%
Netherlands 73%
France 63%
Belgium 45%
UK 36%

The community's rate of growth in its first eight years has averaged 5.6 per cent. Britain's rate over the same period was 3.6 per cent.

Industrial production in the community has gone up by 58 per cent, as against 32 per cent in Britain. The community's exports have increased by 104 per cent, Britain's by only 30 per cent.

Common Market advocates insist that Britain would be able to take a full share of this greater prosperity, that Britain's manufacturers would benefit enormously from being able to offer their goods to an immediate "home" market of 230 million consumers, instead of their present restricted domestic market of 55 million.

Production would be possible along the lines of the American giants (America's home market is 180 million). This could only mean cheaper and better goods in our shops in the long run.

The absence of tariffs within the community would also mean cheaper French perfume, cheaper Brussels lace, Italian shoes, German cameras, Delft china.

But most alarming is what would happen if we don't go into the Common Market.
From the middle of next year all tariff

barriers within the community will be abolished. Imports will be subject to a 22 per cent duty.

Before the community came into existence, Germany operated a 17 per cent duty on imports. So that a £500 car made in France cost £585 to the Germans, as did a £500 car made in Britain. But from July 1, 1968, the French car will cost the German only £500 (the same price as a German-made car) but the British car will cost £610.

Which puts worries about increased food prices in a different dimension.

Letter of the week – Dishonoured price agreement
Robert Fray, Elton, Peterborough 6.2.67

I read with interest your estimate that farm subsidies will be £20 million less than expected. This is only because the Government has not honoured its 1966 prices agreement.

We have received at least 10*s* a cwt less for beef, 6*d* a lb less for mutton, 2*s* a cwt less for barley and wheat, to mention only a few items. They may well have £20 million to spare.

This folly down on the farm could ruin us
John Winter 21.2.67

Home food production is in danger of dropping into the most serious decline since the war. Two years ago it bounded up by 8 per cent. Last year the rise was only 2 per cent. In the year ending next month the rise will be negligible. Next year, unless farmers' confidence is given a massive boost, it will start to shrink – and that's an alarming prospect for all of us.

This is why the annual price review now in progress is the most vital since the reviews started 20 years ago. A run-down in agriculture, leading inevitably to shortages, would be far more serious than any impact of higher food prices caused by Britain joining the Common Market.

The question is whether the Government realises the danger to the industry. If the Whitehall experts who are busy spinning their economic webs in the review talks would pack up for a day and take a trip with me into the countryside, I could soon show them signs that all is not well with agriculture.

There are fewer new tractors working in the fields and fewer new buildings around the farmsteads. At harvest time there will be too many fields carrying indifferent crops of barley, liberally sprinkled with weeds.

These are the result of rank bad husbandry methods into which farmers have been forced by the policy of successive Governments. Continuous cereal production has already led to declining yields and will end in dustbowls if the policy is not changed.

Broken-down fences, unkempt hedges and overgrown headlands are evidence of the continuing squeeze on labour. More and more farmers are reaching the stage where they can no longer substitute machines for men. Farmworkers are leaving the land at the rate of 27,000 a year, far more than the Government thought they would. And they are not all quitting their jobs voluntarily. Some are being sacked by farmers striving desperately to cut costs.

Let the Whitehall experts visit the City and they will find farmers' overdrafts running at the record level of £517 million. This is a 125 per cent increase in ten years. The price of farms sold by auction slumped 12 per cent in the second half of last year. Use of fertilisers dropped last year by 2 per cent – a short-term saving at the expense of long-term fertility.

The most conclusive proof of recession comes from experts who speak the same language as those in Whitehall, the agricultural economists of Nottingham University. Their annual survey of records from 238 farms in the East Midlands (which is usually a good indicator for the country as a whole) shows that average farm income last financial year dropped by 12½ per cent (£428). This was before charging for labour by farmers and their wives, interest on capital or reward for management. The number of farms which made less than £1,000 rose by 28 per cent. Another 9 per cent made less than £500, and 4 per cent showed a loss.

Already the crisis of confidence is making nonsense of the expansion targets for 1970 laid down in the National Plan. The only two major commodities which made significant headway last year were beef and cereals, but in neither case is the build-up in the first one-third of the Plan period anywhere near big enough.

For the rest, milk production is static, sheep down 3 per cent, pigs 10 per cent, egg-laying poultry 4 per cent, potatoes 10 per cent on acreage – with a drop of 900,000 tons (30 per cent) in stocks at the year end.

In the face of such evidence I am baffled by the complacency of Mr Peart, Minister of Agriculture. In the last parliamentary debate on agriculture in November he said: "The figures show we are expanding. The industry is challenging . . . I have every confidence it will continue to progress and fulfil the objectives we have set."

With that sort of thinking at the Ministry, how can the farmers get a review decision which will set production rising?

In a climate of severe restraint, and in the face of Chancellor Callaghan's well-known antipathy to farmers (remember his "farmers at Ascot" gaffe, and the inept handling of the industry in the original Selective Employment Tax plans) it is hard to see how they can expect anything but an austerity settlement which will set agriculture on a downhill rush.

But the National Farmers' Union entered the review in a mood of surprising optimism. I gather it is based on the belief that "Mr Wilson has got the message." Undoubtedly the NFU leaders will take their case to Downing Street if they get no change at the Ministry of Agriculture. It comes down to the question of whether it is wiser to allow an extra few million pounds now, to give farmers the boost they have so far been denied by this Government, or to face immensely bigger sums in a few years to reverse a run-down in production which would undermine the whole national economy.

History would give a damning verdict on the Government and the Prime Minister who started it.

Stansted – It will lose £1m of food every year   Peter Bullen 15.6.67

Land that produces about £1 million worth of food every year will be destroyed if the Stansted airport plan goes ahead. This is the "farm gate" value. In shop terms the figure will be more than double.

The National Farmers' Union put this estimate of the damage that would be done to the economy by Stansted before the Government yesterday. Union experts

and local Ministry of Agriculture officials have spent the past two weeks working it out.

They disagree with the official estimate given at last year's public inquiry that only 2,000–3,000 acres of land will be affected. They say that at least 15,000 acres of the best land in Britain will be taken over. Land that each year produces £730,000 worth of wheat, barley, potatoes and sugar beet; £230,000 worth of pork, bacon, eggs, poultry and beef; £100,000 worth of fruit and vegetables; and about £15,000 worth of miscellaneous crops.

Mr John Walker, Essex NFU secretary, said yesterday: "There is still no detailed plan published, but it is ridiculous to think of only 2,000–3,000 acres. At the public inquiry the discussion was about a two-runway airport. Now they are talking about four runways."

In a formal letter of complaint to the Government the NFU says it has not been appreciated that the airport would not merely absorb many thousands of acres of very high-quality land but must indirectly affect a still larger area. This was because of the enormous volume of extra residential, commercial and industrial development required to service such a major project.

The union called for a fresh survey of alternative sites by independent experts as "surveys undertaken both before and after the public inquiry of alternative sites were inadequate and could not be regarded as objective."

The letter adds: "We share with many other responsible bodies the view that the project would have a devastating effect on the rural scene, upon quality of life over a wide area of the East Anglian countryside, as well as upon the capacity of our land to produce food, upon which the nation must depend."

Stansted Airport – What it will mean to farmers *John Winter 15.6.67*

Mr Cyril Metson, local NFU secretary for 20 years and a close friend of many of the farmers in the area, said yesterday: "In terms of human misery the effects of this plan will be impossible to calculate." Even if only 25,000 acres were taken for the airport about 35 farmers would be affected.

Mr Metson said: "It's not only the farmers and their families, but dozens of farm workers, most near retiring age, whose homes and livelihoods will just disappear."

This is what Stansted will mean to just a few of the farmers:

Mr John Latham, 29, of Waltham Hall Farm, Takeley

He is married with four children under five. The farmhouse was built in 1610. The 312-acre farm is midway between two of the proposed runways. He will lose it all.

Annual output: potatoes, 1,200 tons; wheat, 180 tons; barley, 170 tons. Labour: four men, all over 55, some with nearly 40 years' service.

Mr Geoffrey Brown, 56, High House Farm, Takeley

He is married with two daughters and has farmed the 168-acre High House for 46 years. He will lose it all.

Annual output: wheat, 120 tons; potatoes, 120 tons; beans, 16 tons; peas, five tons; silage, 100 tons; grass seed, two tons; 30 beef steers. Labour: two men, with 15 years' service, and a boy.

Mrs Nancy Ritchie, 29, Motts Hall Farm, Elsenham

Mrs Ritchie, married with two sons, started a pig-rearing unit nine months ago, partly with £2,000 saved by herself and her husband. The rest was borrowed. The farm is near the end of the present Stansted runway. She will lose it all.

Annual output: 500 quality pork pigs.

Mrs Ritchie said: "What is going to happen? What compensation will we get for a house and nine acres? I do not imagine it will even be enough to pay back what we had to borrow."

Jets already rev up 200 yards from their windows. She said: "The noise is absolutely indescribable, although it is only an occasional plane at the moment."

The threat of Stansted prevented them selling their house some time ago when her husband wanted to move with his firm. He was out of work for six months as a result.

Mr Geoffrey Gowlett, 48, Thremhall Priory Farm, Gt Canfield

The 250-acre farm is at one end of the Stansted runway. He will lose all but ten acres.

Annual output: barley, 160 tons; wheat, 200 tons; sugar beet, 800 tons; beans, 70 tons. Labour: three men with 15 years' service each.

Mr Gowlett said: "I have been farming here since 1932 and my family have farmed in the district for generations."

Mr John Pimolett, 54, Tye Green Farm, Tye Green

The farm is 460 acres. He will probably lose it all.

Annual output: barley, 625 tons; wheat, 255 tons; potatoes, 300 tons; hay, 75 tons; 30 beef steers. Labour: seven men, one with 40 years' service, the others about 30 years.

Mr Pimolett would also lose his 16th-century farmhouse. He said: "I have been farming since I was 16. At my age a man doesn't want to start all over again."

CROPS and MACHINERY

Save our soil, says farmers' leader
John Winter 4.1.67

The price squeeze imposed by successive Governments is robbing British farmlands of fertility built up over a century, one of Britain's best-known farmers said yesterday. Mr Charles Jarvis, chairman of the British Farm Produce Council, told 900 delegates at the Oxford Farming Conference that within the next 20 years Britain would be faced with food shortages if Government policy was not changed. Fears that dictation from Whitehall is mortgaging the future to pay for present production were widely expressed.

Mr Jarvis, who farms 570 acres at Little Clacton, Essex, said that continual Government bullying, abetted by the promptings of its own National Agricultural Advisory Service, had forced farmers into methods which were little short of disgraceful. The "gross margin" formula had become the criterion of a farmer's success instead of the real test of good husbandry.

"Over the next decade or two Governments will have to face the prospect of too many people on this too small island not having enough to eat. This island is in a position of great peril unless the Government gets firmly into its thinking the importance of conserving the soil and building up fertility, instead of splitting

it as it is doing at the moment."

The major threat to the future, Mr Jarvis told me, is the continuous growing of cereal crops on the same fields. Barley or wheat had been grown without break for ten years or more on a huge acreage of land. Two or at the most three successive crops were the maximum in a good husbandry system.

Professor Dennis Britton, head of Nottingham University Department of Agricultural Economics, said: "We have pressed so much for quantity of yields that we cannot sustain it. New diseases and weeds are building up."

Record machinery exports
John Winter 4.1.67

Exports by farm machinery makers earned £166,300,000 between January and November last year, £13,700,000 more than the same period in 1965. For the whole year exports are expected to reach a record £180 million.

Weed sprays can cut corn yields
Peter Bullen 13.1.67

Farmers were warned yesterday of evidence that weed killers were not increasing the cereal yield and in some cases were actually reducing it. Mr Emrys Jones, the first chief agricultural advisor to the Ministry of Agriculture, told more than 300 farmers at Stoneleigh,

Warwickshire, that the Ministry's advisory service was making a big check on the use of chemical weed killers.

The 1965 survey had shown that in many cases there was either no benefit to cereal yields or there was a drop in yields. But in most cases where yields went down it was as a result of "slap-happy" spraying by the farmer.

Mr Stanley Evans, the Ministry's liaison officer at the Weed Research Organisation near Oxford, said last year's results looked like confirming the 1965 findings. Although the sprays might not increase yields they certainly cleared up the weeds, allowing easier harvesting. But the chemicals had to be applied accurately.

The cereal producer alone had 300 different chemicals from which to choose. By choosing the dearest treatment as opposed to the cheapest he could cut his net profit by 25 per cent an acre, assuming yields were the same for both treatments.

Letter of the week – water bills

Alfred Dring, Towcester, Northants 30.1.67

I have two bills on my desk. One is for a payment to the drainage board to put water in my brook; the other to the river authority to take water out.

Note: Under current legislation a drainage board levies a rate on farm land for drainage which results in water being carried away by a brook. A farmer is required to pay a licence fee to the local river board for water which he takes out of the brook for irrigation.

Frost helps wheat   *Peter Bullen 15.2.67*

Farmers all over the country are making desperate last-minute efforts to sow their winter wheat. The past week or so of dry, frosty and windy weather has at last started to dry out soil which has been sodden and unworkable for months. In the country's main corn-growing area of the Eastern counties, ploughing and sowing has been going on until nine and ten o'clock at night by the light of tractor headlights. Only about two more weeks are left during which winter corn can be sown.

The prospect of less wheat and more barley being produced this year has caused a lot of concern to the Minister of Agriculture, Mr Fred Peart, and the Home Grown Cereals Authority. Last year wheat production dropped, while barley output rose to a record 8.8 million tons. This prompted Mr Peart to give a broad hint that in the farm price review talks now taking place he will take steps to put a damper on barley production and do something to pep-up wheat output.

David Brown Selectamatic

John Winter 7.3.67

Biggest and most powerful David Brown tractor makes its first public appearance today at the international agricultural machinery exhibition in Paris. It is the 67 bhp 1200 Selectamatic, weighing over 2½ tons.

Safety cab curb on field deaths

Roy Gregor 1.6.67

Regulations to cut farm casualties were announced by the Government yesterday. From September 1, 1970, all new tractors will have to be fitted with safety cabs or frames and within ten years all tractors.

In the last ten years, 423 farm workers have been killed by overturning tractors. Last year alone this was the cause of 53 out of the 135 fatal accidents on farms. The Ministry of Agriculture said yesterday: "Investigations have shown that most, if not all, of the drivers' lives would have been saved if they had been protected."

The new regulations will also ban a farm worker from driving a tractor that is not equipped with safety cabs or frames. The only exemptions will be for work in orchards and awkward buildings. The cost is expected to be between £50 and £100 a tractor.

A Scottish Farmers' Union official said: "We welcome any measures that contribute to safety on farms and the ten year period is reasonable."

Why farmers should follow the plough   Peter Bullen 7.4.67

Ploughmen of England should revert to the centuries-old custom of following the plough instead of pulling it behind a tractor, says a professor of ergonomics.

Ergonomics – the study of man at work – is revealing glaring weaknesses in machinery designs which put unnecessary stress and strain on workers, impairing efficiency and even health. One example spotlighted by Professor W F Floyd, Professor of Ergonomics and Cybernetics at the University of Technology, Loughborough, is tractor driving.

In this month's *Design*, the journal of the Council of Industrial Design, he says: "All too often the tractor driver is expected to lean or twist backwards to operate the machinery attached or trailed behind him. A driver should sit at the back of the power unit with the tools and implements mounted in front of him on a tool bar." This would give the driver a good view of his work task ahead, steer-

ing would be easier, and all controls would be in front of him, ergonomically correct in design and readily accessible.

Professor Floyd has many criticisms of agricultural operations and equipment. Writing for the Royal Agricultural Society of England's journal he says there is some evidence of deafness in the agricultural population from the noise of tractors and other machinery.

This month he will be leading speaker at a three-day conference, organised jointly by the Royal Agricultural Society and the Ergonomics Research Society, at Nottingham University's School of Agriculture. The conference, from April 19 to 21, will cover all aspects of ergonomics in farming and will be the first on the subject in Britain. It is a triumph for the National Union of Agricultural Workers, which has pioneered ergonomics research in farming over the past two years.

Boom harvest hits grain prices
                              John Winter 23.8.67

A flood of grain from Britain's sundrenched harvest fields is threatening to swamp the market and undermine prices. Corn-merchants are appealing to farmers to keep the grain on their farms until the glut has been absorbed.

If farmers insist on getting rid of their barley now they will receive a poor price, and this in turn will push up the subsidy which the Government has to pay to meet the price guarantee. Already the price for barley for livestock feed has slipped by 30s a ton since the harvest started.

Mr Gordon Wood, of Cambridge, president of the National Association of Corn and Agricultural Merchants, urged farmers and merchants to fill all their available storage capacity in an effort to hold grain off the market and avoid any unnecessary fall in price. "The heat-wave could not have arrived at a more opportune moment, as cereal crops were ready to harvest throughout England and Wales," he said.

But, warned Mr Wood, the good weather could cause grave problems for merchants and manufacturers if grain for immediate delivery is pressed on them. Grain harvested in the heat-wave carried such a low moisture content that, carefully stored, it could be kept up to two months on the farm without harm. By that time manufacturers would be ready to take further supplies.

The National Farmers' Union backed Mr Wood's appeal. Mr Peter Savory, chairman of the union's cereal commit-

tee, said: "This is sound advice. There should be a good demand for grain throughout the year. Given proper marketing, there is no justification for a price fall."

A record 8,067,000 acres of corn have been grown in England and Wales this year, 105,000 acres more than last year. This will bring the UK total to nearly 9.5 million acres. Yields are running on average more than 1 cwt an acre above last year. With good weather for the rest of the harvest, a record total crop of around 14 million tons of wheat, barley and oats will be gathered.

This would be worth £280 million at market prices, excluding subsidy. But forced selling in the next few weeks could chop millions off its value.

 # LIVESTOCK and POULTRY

Deborah will rear a very special calf
John Winter 6.1.67

First fruit of the French invasion of Britain's pedigree cattle business has arrived on a farm in Oxfordshire – the first pure Charollais calf bred in this country.

The calf, a sturdy, 72lb bull with a creamy white coat, has been bred by Mr Roger Chapman and his wife, and is being reared by their daughter Deborah, aged 14.

Their farm at Mapledurham, near Reading, covers only 27 acres, and their total herd is five Charollais heifers, part of the first importation into Britain a year ago, and one or two non-pedigree beef animals.

Cattle disease hits four counties
John Winter 9.1.67

A foot-and-mouth restriction order was placed on West Sussex, Hampshire and parts of Wiltshire and Dorset at midnight, banning the movement of livestock in the area.

Last night the fourth case of the disease in 48 hours was confirmed on the farm of Mr Arthur Jeram at Bedhampton, Hampshire. Several hundred pigs will have to be killed.

The outbreak began on a farm at Southwick, near Fareham, Hampshire, on Friday. Two other cases confirmed on Saturday brought the total of animals slaughtered then to 204 cattle and 60 pigs. All the stricken farms are within five miles of the first outbreak, but the quick spread of the disease caused the Ministry of Agriculture to send veterinary staff from a wide area at the weekend.

By last night more than 20 vets were based at a disease control centre in the Territorial Army centre at Fareham. Inspection of farm animals throughout the area will continue today.

Foster breeding triumph in sight
John Winter 11.1.67

Farm animals can become mothers of offspring conceived by other females and transferred to the foster-mother as fertile eggs. Scientists at Cambridge have perfected the transplantation technique in sheep and pigs, and experiments with cattle are progressing.

Mr Tim Rowson, deputy director of the Agricultural Research Council's breeding research unit told the British Cattle Breeders' Club conference at Cambridge yesterday that the success of recent experiments with cattle there will have a "snowball effect" as scientists in other countries take up the work.

The main value of transplantation is to enable a breeder to get many more offspring from outstanding animals than could be bred by natural means. Mr Rowson visualised the time when artificial insemination centres would keep stocks of fertilised eggs on ice, which could be implanted into cows of inferior breeding potential. They would give birth to superior calves which would be unrelated to them.

Foot-and-mouth signs brighter
Peter Bullen 23.1.67

With only one new case in the past five days the Hampshire foot-and-mouth

disease outbreak could be dying out. But the huge team of nearly 50 Ministry of Agriculture vets are still on the alert in the disease-control headquarters at Bishops Waltham Civil Defence centre.

This is the stage they fear most when hopes are beginning to rise, but any minute could bring an alarm call to a farm outside the ten-mile infected area round Fareham in which they have managed to contain all the 27 cases so far.

If there are no further outbreaks some of the restrictions might be relaxed in a few days' time and some of the vets stood down. Since the first outbreak on January 6, 2,682 cattle, 4,727 pigs, 386 sheep and six goats have had to be slaughtered. Compensation for them will cost more than £270,000.

A 15,000-word report on last year's big foot-and-mouth outbreak in Northumberland was sent to the National Farmers' Union headquarters this weekend by its Northumberland branch. But the report is not the "dynamite" that had been expected and is unlikely to lead to more pressure on the Minister of Agriculture, Mr Fred Peart, to hold an independent inquiry.

Egg board wants to drop the Little Lion
John Winter 30.1.67

It looks as though the Little Lion will soon be gone. The Egg Marketing Board, which has used him as a brand mark for ten years, has decided that the only way to meet the growing challenge of unstamped eggs is to kill him off.

Egg packers, the board's agents, who stamp the lion on the shell, say that the number of eggs diverted from their stations for sale unstamped at higher prices is rising continually, and the argument is: "If you can't beat them, join them." The board is also supported by most other sections of the industry and would like to see the lion's death warrant signed at next month's price review.

But the National Farmers' Union is not convinced that killing the lion would be in the best interests of all of its members. If all eggs, including the 63 per cent which still pass the board's testing and grading standards at 454 packing stations, were sold unstamped, the extra 6d to 1s a dozen which shoppers pay for unstamped, untested eggs at farm gates would fall sharply.

The number sent to packing stations would increase, but the price would not rise by the same amount. An NFU official said: "We have got to try to assess the effect of the removal of the lion on the total return from eggs to producers. Abolishing the stamp is not necessarily the solution to the sale of eggs away from the board."

The board, which has been spending £1.6 million a year on promoting lion eggs, admits that its share of total home production is steadily shrinking. The board says: "Marketing costs, increased by selective employment tax and other causes, are now greater than the egg subsidy, about 6½d per dozen this year. This creates an inducement to producers to sell their eggs away from the board if they can get a better price."

Villagers complain of smell
Peter Bullen 14.3.67

A new £100,000 plant which turns manure from egg and chicken factory farms into processed fertiliser has run into trouble. Villagers at Methwold, Norfolk, complain that fumes from the plant's smoke stack are making them ill. Mr Paul Hawkins, MP for South-West Norfolk, has taken up the matter with county and local government authorities and villagers are holding a protest meeting tonight.

The multi-million Ross Group, which built the plant, called in gas purification experts yesterday to plan a series of screens and water filters to extract obnoxious vapour and particles from steam rising from the 75ft chimney.

Mr Alex Alexander, 50, chairman of the group's poultry division, said: "We have had a number of teething troubles, including the smell. We have solved the other problems and we shall solve this one. We want to establish good relations with our neighbours."

Cattle with horns will be no more
John Winter 29.3.67

Horned cattle will have almost disappeared from British farms within two years when regulations will have compelled farmers either to breed from naturally horn-less strains or to de-horn calves within days of birth.

Reason for the campaign is that horned cattle do considerable damage, especially to each other, and they require more space in markets and in transit. Injuries which a horned animal can inflict on another often result in permanent damage to the hide, reducing its value as leather. In serious cases, damage to meat is also revealed after slaughter. In horned dairy herds cows are sometimes seriously injured around the udder in skirmishes.

The most decisive anti-horn move has just been made in Ireland. The Governments of both Eire and Ulster have announced a total ban on horned cattle for sale or export after February 1, 1969. More than 500,000 live cattle were shipped to Britain from Ireland last year. As the number should rise to more than 800,000 by 1969, the effect of the ban on our own cattle population will be immense.

Although our Government does not contemplate banning horns, regulations are being prepared to enforce the segregation of horned and hornless cattle in road wagons, railway vans, ships and aircraft. The Agriculture Ministry believes that these measures will have the same effect as the Irish move, without an actual ban on horns which would be opposed by some cattle breed societies.

Foot-and-mouth 'should be national emergency'
Roy Gregor 22.4.67

Foot-and-mouth disease outbreaks, a group of farmers suggest, should be treated as a national emergency with the Army and Civil Defence called in. That is the main recommendation of a report compiled by leaders of the Northumberland and Roxburghshire areas of the Farmers' Union. It will be sent to the Minister of Agriculture.

Last year Border farmers lost 45,000 animals when the disease swept across Northumberland. Farmers from both sides of the Border are extremely critical of the handling of the outbreak, which cost the Government £900,000. They complain about a lack of warning to farmers in the area, lack of speed in taking action and inefficient slaughtering and disposal of animals.

"We believe the greatest single requirement for combating an outbreak is that the Government should regard the matter as being an emergency," states the report.

Bid to wipe out 'vet disease'
John Winter 22.4.67

First step towards wiping out brucellosis, the cattle disease which can also be transmitted to humans, starts today. The Ministry of Agriculture is ready to receive applications from farmers for their herds to be registered as disease-free. It will be at least nine months before the first herds can complete a series of four tests which will establish that all the animals are clear.

The disease, which causes cows to abort, is estimated to cost dairy farmers £1 million a year in loss of milk and calves, and is a hazard to people who have contact with cattle. Nearly two-thirds of vets tested last year showed evidence of infection, and 27 per cent had symptoms of chronic brucellosis. People can also catch the disease (known as undulant fever) through drinking milk from infected cows unless it is pasteurised.

Not until a substantial reserve of disease-free cattle have been registered will the Ministry launch compulsory eradication, with compensation for slaughter of infected animals. This would not be for two to three years.

The British Veterinary Association, which has campaigned for action, last night welcomed the move but stressed that it is only a preliminary. An immediate eradication scheme, says the Ministry, is impracticable because an estimated 14 per cent of cows would react to tests. Costs would be about £40 million. A progressive scheme would cost only about £10 million, spread over 15 years.

Pigs are put on the pill
John Winter 28.4.67

Family planning for pigs is now being tried out at more than 200 commercial farms. The pigs are given a drug called methallibure in their food for 20 days. The drug postpones the breeding condition, and when the dose ends, all the animals are ready for mating at the same time.

The case for synchronising mating – and the birth of litters – is the difficulty of detecting the proper breeding condition in pigs. By planning the mating, birth and weaning of the whole herd in advance, the farmer can reduce the loss of piglets – now running at 25 per cent. And if the new technique is widely adopted, it could overcome the disastrous drop of more than a million pigs for slaughter, which has nearly crippled the British bacon industry.

Mr Tom Groves, veterinary surgeon in charge of field trials, said: "The cost of treatment is less than half the £3 it costs for an inseminator to visit a farm to treat perhaps only one pig which might not be in correct breeding condition."

Methallibure's ability to control the breeding cycle was discovered by ICI scientists at Alderley Park, Cheshire.

Life is champion thanks to Moreta
Peter Bullen 15.5.67

The Royal Smithfield Show has dropped its rule that the supreme champion must be slaughtered because of the heifer that was too beautiful to die. Pride Moreta of Thorn, the supreme champion last year, should have been killed according to the show's old regulations.

But there was such an outburst from animal lovers that Moreta's owner, Mr John Evans, 45, of Cardigan, refused to kill her. She is now being kept by a farmer in Cardigan at a cost of £3 a week.

By dropping the slaughtering rule the show organisers hope to avoid public outbursts of sympathy so that butchers who buy the champion will not be under pressure from their customers.

Beef-hungry Russians buy 1,000 white faces
John Winter 6.7.67

The biggest export deal of the century in pedigree cattle was being completed with Russia at the record-breaking Royal Show last night. The Russians are ordering 1,000 white-faced Hereford beef cattle to be delivered over the next five years. The deal is worth at least £200,000.

The cattle, mostly females, but including some bulls, will be selected by the Hereford Herd Book Society. They will be quartered on two new breeding farms in Russia and the society will advise on breeding programmes.

The Russians, facing a chronic shortage of quality beef, have found that the fleshy, fast-growing Hereford produces an excellent beef calf when crossed with one of their native breeds, the Kazen.

The first export to Russia of three white-faces was in 1910. Since the war about 1,000 head have gone out in batches including the 23 which represented the Hereford breed at the British Agricultural Exhibition in Moscow in 1964.

The Russians estimate they have 2 million Hereford crossbreds throughout the country. Large-scale production on a planned breeding programme will quickly build up the numbers and quality of their beef herds.

The order is double that for Spain negotiated in April by Mr Tony Morrison, secretary of the Hereford Society, and it will be the biggest export consignment since 3,000 were shipped to Oregon in 1885.

Peart picks a team to investigate the Little Lion

Peter Bullen 27.7.67

A Government commission is to be set up to consider the future of the Egg Marketing Board and its Little Lion stamp. It will also investigate all Britain's supplies of eggs and egg products, both home-produced and imported, and it will report back early next year.

Mr Fred Peart, Minister of Agriculture, announced this yesterday in the Commons, where he was under attack at Question Time about the chaos in both the egg and beef markets. Farmers' prices for beef and eggs have slumped drastically because of high home production aggravated by uncontrolled imports, some of which have been "dumped" on the British market at artificially low prices.

Mr Peart said that urgent discussions between the Government and the egg industry would continue on questions such as future supplies, including imports, and modifications to the Egg Board's producer contracts scheme. But other far-reaching proposals have been put forward which would involve a fundamental reappraisal of the marketing arrangements and he had decided to set up a reorganisation commission under the Agricultural Marketing Act, 1958.

It would cover all aspects of egg marketing, including the 1956 scheme under which the Egg Board was set up.

Plain old Rose wins top dairy show award

John Winter 25.10.67

Princess Margaret kept her fingers crossed yesterday for the Guernsey cow called Stype Rose 13th. And ten-year-old Rose pulled off the most sensational and popular win in the history of the Royal International Dairy Show. She beat smarter rivals half her age to become supreme individual champion.

Her owners, Mr Alec Vincent and his wife, Nancy, hugged each other with joy as Princess Margaret stroked the cow. Herdsmen flung their hats in the air and applause echoed through Olympia, London. Many remembered how matronly Rose, with a big lead on milk points at last year's show, was beaten for the title on looks.

The Princess said to the Vincents: "I told you I would keep my fingers crossed for you, and I did." When Mr Vincent said that Rose was not considered very good looking, the Princess replied: "She looks all right to me."

The judge, Mr Herbert Tully, put Rose tenth in line on appearance with 88 points out of a possible 125. With her record total of 280 points for 10.7 gallons of milk she gave in the show trials last Saturday no animal among the 14 in the ring could touch her. She finished 42 ahead of the runner-up, a five-year-old British Friesian, Suttonhoo Dividend Pietje, owned by Mrs Anne Barton, of Woodbridge, Suffolk.

The Vincents have only 54 acres on their New Forest farm at Blashford, near Ringwood, Hampshire, but they keep 70 Guernseys.

Disease corrals show cattle

John Winter 26.10.67

Over 300 animals will have to be destroyed after a mystery outbreak of foot-and-mouth disease at Mr Richard Ellis's Bryn Farm at Nantmawr, Oswestry, Shropshire, yesterday. The disease was confirmed in a pig, and the movement of animals has been banned in a five-mile radius of the farm.

Mr Ellis said last night: "How the outbreak started is a complete mystery and it is likely to take a long time before it is discovered. It is a dreadful loss of animals."

And later cattle from the five counties adjoining Shropshire were forbidden to leave the Royal International Dairy Show in London yesterday. All the animals concerned were detained until they had been examined and clearance certificates issued. All were found to be free of disease.

Hundreds of cattle, sheep, and pigs were impounded in the open market at Oswestry and will have to stay there until the ban is lifted. Police sealed off exits to the town and turned back cattle lorries which had left the market before the ban was announced.

Sweeping foot-and-mouth baffles experts   *John Winter 2.11.67*

Foot-and-mouth disease is spreading quickly across Britain – and baffling experts from the Ministry of Agriculture. Outbreaks in the Welsh border country, Cheshire and Lancashire, have doubled within 24 hours.

Last night, the total of infected farms had risen to 37. The development of the disease within a week of the first outbreak near Oswestry, Shropshire, is more widespread than any in recent years.

Mr Fred Peart, the Minister of Agriculture, said last night: "The situation is serious and it could become *very* serious."

The 32 confirmed cases equal the total in last year's disastrous epidemic in Northumberland, but have occurred six times as quickly.

SOS for vets as cattle menace spreads   *Peter Bullen 14.11.67*

Britain is having its worst year for foot-and-mouth disease for 15 years. And the epidemic continued to spread yesterday. Private veterinary surgeons are to be asked to help the 150 Government and foreign vets who are working almost round the clock to fight the disease.

More than 50 new cases were confirmed over the weekend bringing the total to 272 and with earlier outbreaks in Hampshire and Warwickshire will make 1967 the worst year since 1952. So far this year 27,067 cattle, 15,648 sheep and 19,654 pigs have been slaughtered. Compensation payments alone total almost £2½ million.

In 1952 there were 495 cases. In 1942 there were 670.

Ministry of Agriculture vets stressed yesterday that although the present epidemic is serious it is by no means a national disaster yet. The number of animals affected is only a very small fraction of the country's total livestock population, which is nearly 50 million animals.

But the pattern of the disease took a turn for the worst yesterday. Five cases were confirmed in Montgomeryshire and Denbighshire and one near Kendal, Westmorland, the first outbreak in the county since the start of the epidemic. For the first time the disease is spreading back or westwards away from the original outbreaks near Oswestry in Shropshire. Until now the prevailing winds seem to have spread the disease thickly over the ground towards the east.

From the main disease control HQ in Oswestry last night Mr Tom Stobo, the Ministry's regional veterinary officer for Yorkshire and Lancashire, advised farmers on the hills to house any cattle they can. Farmers should try to gather their sheep on the inner part of their farms to leave as wide a cordon of land as possible around them.

The Ministry of Agriculture said yesterday: "We have been unable to put our finger on the precise source of the disease, although we are confident it entered Britain in imported meat."

Ghouls flock to cattle killing county
Peter Bullen 14.11.67

Car-loads of sightseers have been blocking country lanes in the heart of the foot-and-mouth disease area in Cheshire. A farmer said last night: "These ghouls want to see the animals being slaughtered and burned or buried. It's despicable."

The disease control headquarters in Oswestry said: "It is alarming as the disease is spreading particularly rapidly, and any movement of people or vehicles in the infected areas increases the risk of further spread."

The NFU has cancelled the monthly meeting of its general purposes committee and council tomorrow and Thursday. An official said: "We felt it would be wrong to bring farmers from all over the country to a meeting at this time."

The RAC has tried to find ways of re-routing its international car rally to avoid disease areas. An official said: "All special stages west of a line from Bristol to Carlisle are cancelled. We are now faced with the problem of getting to Scotland and back without going over these affected areas."

The disease has now been confirmed in eight counties: Shropshire, Cheshire, Derbyshire, Lancashire, Flintshire, Denbighshire, Montgomeryshire, and Staffordshire.

Silent disaster: only an empty farmyard is left to mark the hard work of 15 years
John Winter 17.11.67

Three weeks ago it was a bustling farmyard. Now the only animal left is Nell the collie. For days she has been sad and silent. She misses the 50 British Friesians that made up the pedigree Sevenwells herd, once the pride and joy of Harold and Sylvia Jones.

Three weeks ago the Joneses watched a heifer being born at their farm at Nantmawr, Shropshire. Sylvia, 35, said: "It was really exciting. The little thing was the first calf by a bull we bought when he was six days old for 100 guineas." Then the symptoms of foot-and-mouth disease began to show in the mother. Their world at Prospect Farm collapsed. First the vet came, then the valuers, then the slaughterers, then the burial gang, then finally the mopping-up squad.

Harold Jones, 43, started the herd with three heifers 15 years ago. He built it up until it stood high in the county milk records. Outside the 50-acre farm still swings the sign of the British Friesian Cattle Society, proclaiming that the herd was among the elite. Recently the herd was yielding 80 gallons a day, bringing a £365 milk cheque every other month. Now that has stopped. Government compensation covers only the market value of the animals.

Harold said: "You just cannot replace a herd like that by going out and buying cows. You've got to breed them, and the time and planning matter most. Meantime, we've got to live."

Sylvia is devoting more time than usual to the children, Jennifer, aged seven, and five-year-old Alan. She said:

"They loved the cows too. If we were half an hour late getting home the whole herd was watching for us at the gate and the calves were bawling their heads off. Now it's so quiet it's frightening."

Too little, too late, say Tories on cattle killer John Winter 4.12.67

Minister of Agriculture Mr Fred Peart faces severe criticism in the Commons today over his handling of the country's worst foot-and-mouth epidemic. The attack, during the agricultural debate, will be led by Mr Joseph Godber (Tory, Grantham, Lincs), the shadow Minister of Agriculture.

He will accuse Mr Peart of misplaced optimism in the early stages of the outbreak and then of doing too little, too late. Opposition MPs will demand a clear statement over imports of meat from countries with foot-and-mouth diseases – suspected source of the British outbreak. This follows Friday's reports that a ban on meat from South America was first imminent, and then not so imminent.

Farmers are angry over the Ministry's advice that disinfectant pads and splashes are unnecessary on main roads far from infected areas. Mr Airey Neave (Tory, Abingdon, Berks) has sent a letter to the Prime Minister protesting at the withdrawal of troops who had manned disinfectant barriers on 15 bridges over the Thames.

Mr Peart will also be accused of failing to make sensible preparations for a switch of policy from slaughter of animals on infected farms to vaccination. Farmers are further annoyed over Mr Peart's statements last week that there was no reason to panic about introducing special measures to stimulate extra home production. Mr Godber will claim that farmers could do more than anyone else to close our trade deficit.

Letters of the week – Could aircraft be to blame? 4.12.67

Dozens of readers have written letters about the foot-and-mouth disease epidemic. The majority attack the Government's slaughter policy. Others are still suggesting possible causes for the original infection:

C Marston, Chiswick, London

The infection could be brought in by aircraft. It appears to me that the areas affected are mainly those over which aircraft fly.

"Country Lover", Manchester

What about all the flying saucers seen around just before the epidemic?

Lionel Arculus, Eckington, Worcs

I have written often to the Ministry of Agriculture pointing out the futility of the slaughter policy. I am 71, and if I were farming today the powers-that-be would never carry out their devilish policy on my place.

Mrs Jones farms by telephone

John Winter 5.12.67

Although she has only lived on a farm for a year, doctor's wife Mrs Jenny Jones has had to take a telephone course in instant cow care because of the foot-and-mouth epidemic. When the disease started to devastate Shropshire farms, six Friesians were "frozen" in the field owned by the Joneses where they had grazed all summer. They belonged to a farmer whose land, six miles away, was struck by the disease.

So, Mrs Jones, of Llan Farm, Llanyblodwell, had to look after them.

And the former nursing sister has managed very well. Already she has delivered two calves, Lennie and Jennie, after being briefed by the animals' owner, over the phone. She said: "I did not do midwifery, and I was a bit out of my depth."

Now, the cattle on the farm are the only ones alive in the Tanat valley.

Egg men sack board chairman

Peter Bullen 7.12.67

Egg producers showed their dissatisfaction with the Egg Marketing Board yesterday by throwing its chairman, Mr Christopher Harrisson, off the board.

Mr Harrisson, chairman for 4½ years, lost his seat when he came second, with 10,200 votes, in the poll for the East Midlands Regional member. He was defeated by an "unknown" producer, Mr Martin Middlebrook, who polled 28,029 votes. It was the first time Mr Harrisson had been opposed in a board election since 1957.

Mr Middlebrook, 35, is a commercial egg producer with 9,000 hens and two pullet rearing farms at Wyberton, Boston, Lincolnshire.

The chairman of the British Egg Association, Mr Ray Feltwell, said the election should not be interpreted on a personal basis. "It is clear that egg producers of all sizes took part in this protest vote and in so doing clearly demanded radical changes in egg marketing arrangements."

The defeated Mr Harrisson said last night that he did not know yet whether he would stand again for the board. He would carry out his duties until December 31 when he leaves.

Board members will meet on January 3 to elect a chairman. Most likely candidate is the present vice-chairman, Mr Geoff Kidner.

Heartbreak day as village loses all its animals

John Winter 13.12.67

Every cow, sheep and pig in the tiny Lincolnshire village of South Hykeham (pop. 60) was condemned to slaughter yesterday when foot-and-mouth disease made its third strike in the county. Even a farm hand lost the five pigs he kept in his back-garden sty.

The day of heartbreak started with farmer John Phillips, 62, noticing one of his pedigree Dairy Shorthorn herd limping. The village was soon ringed by police road blocks. Only essentials such as food and mail were allowed through. No one from the farms was allowed out. The vicar, the Rev Frank Sargeant, was told he would not be allowed through from a neighbouring parish to conduct services.

Thirty more cases were confirmed yesterday. Mr Fred Peart, Minister of Agriculture, said in a written answer in the Commons that 1.2 per cent of the 12.3 million cattle in the United Kingdom had been slaughtered, 0.2 per cent of the 28.9 million sheep and 1.1 per cent of the 7.1 million pigs.

The Irish Ambassador, Mr John Molloy, a widower, said he would not be going home this year to spend Christmas with his sons Tom, 12, and Jack, 11. And on the Christmas cards he sends home will be written: "Read and destroy". It is feared that mail might transmit the disease.

Children stranded

John Winter 20.12.67

Sixteen children and about 14 adults will

have to spend Christmas stranded on a farm which has been hit by foot-and-mouth – attached to Kesteven Agricultural College, near Caythorpe, Lincolnshire. They are not expected to be allowed off the farm until after the New Year.

Cattle plague may be over in new year
Peter Bullen 27.12.67

Hopes that the end of the foot-and-mouth epidemic will be in sight in the first months of the new year continued to rise yesterday following a quiet Christmas period. The epidemic continued its slow decline over the four-day period and the Christmas Day figures of 16 new cases was the lowest since November 6.

The number of confirmed cases up to 5 pm each day was: December 23, 20; Christmas Eve, 20; Christmas Day, 16; and yesterday at 8.15 pm, 21. Large parts of the disease area have already been released from infected area restrictions in the past few days. Further reductions around Lincoln are expected tonight.

But Government vets are disappointed by the continued spread in Staffordshire and they are also worried about the number of new cases occurring in sheep. On Christmas Eve nearly 3,000 sheep were condemned. Since then 1,160 more have been slaughtered.

Altogether 14,642 cattle, sheep and pigs were destroyed over the Christmas period, bringing the total in the epidemic to 173,331 cattle, 75,945 sheep, 94,689 pigs and 23 goats. Compensation payments are now near the £19.5 million mark.

EMPLOYMENT

To be, or not to be, a farmer's boy
John Winter 5.1.67

To be a farmer's boy today means to the average school leaver being paid low wages for a dirty job with bad amenities. Leading farmers admitted this at the Oxford Farming Conference yesterday. They warned that unless a new image is found for the land there will soon be no farmers' boys either to plough and sow or work the automated farms of the future.

Mr John Pollard, principal of Berkshire Institute of Agriculture, said that between 1963 and 1966 the number of farm boys under 20 had fallen by 26.5 per cent and those under 18 by 32 per cent. Among older men the drop was only 5.7 per cent.

Mr Pollard said: "It is vital that we outline the advantages of joining agriculture. There is no such thing today as an agricultural labourer. The term 'unit manager' is often highly appropriate and is a step up the promotion ladder."

Mr Ted Owens, a Somerset farmer, said: "We have got to use shock tactics to clean up our image. We must hit the headlines with a big plus rate for the skilled operative and wage structure based on the ability of the man."

Letter of the week – farmworkers will not stand much more
P Gibbons, Seabrook, Hythe, Kent 9.1.67

It is a public scandal that the steadily dwindling number of highly responsible farm workers should be publicly humiliated each time they ask for a pay rise. The decision to refer the latest 6s-a-week award to the Prices and Incomes Board (denounced as a "smack in the eye" by a leading farmer at the Oxford Farming Conference) is, for me, the last straw.

Let Mr Wilson understand that farmworkers will not stand for much more of this nonsense meted out by the Agricultural Wages Board and the Government.

NFU plans bonus for skilled farm men
John Winter 25.1.67

The National Farmers' Union plan for higher pay rates for skilled farmworkers was announced at the annual meeting in London yesterday by Mr Henry Sharpley, chairman of the labour committee.

Three levels of premium above the minimum wage (which is expected to be raised to £10 16s, from February 6) would be related to individual responsibility of workers.

The first level would be paid to men who could establish that they were skilled. If a worker of three years' experience could not agree this with his employer, he could establish his right by passing proficiency tests. Second and third levels would be based on responsibility as well as skill and experience and would depend on the workers being appointed to a responsible position by the farmer. The premiums would be percentage additions to the basic wage.

Mr Sharpley said: "It is important not to increase the possibility of demarcation disputes, which bedevil other industries, nor jeopardise the excellent labour relations usually found on farms. The structure must also apply to the vast range of sizes and types of farms which come under the jurisdiction of the Agricultural Wages Board, and it must be simple to administer."

Talks on the new deal start at the board next week between representatives of the NFU and the Agricultural Workers' Union, which favours a straight rate-for-the-job system.

Letter of the week – Mother, I've seen a cabbage!
E F Woodward, Hextable, near Swanley, Kent 6.3.67

THE farmworker looks over the fence and sees the large army of non-productive workers receiving higher wages and higher pensions on an earlier retirement. The ugly building monster is spewing out concrete and bricks all over our valuable farm land.

The day is near at hand when the little boy will come home from school and say: "Mother, I've seen a cabbage."

Put teacher on the road, says board
John Winter 10.3.67

Mobile classrooms may teach the latest mechanised farming techniques to men working alone on small farms. Alternatively farmers could draw on pools of substitutes while their one worker attended a course. These are among suggestions by the new Agricultural Training Board to help farmers who employ only one man and cannot afford to release him for training.

The Board estimates that half Britain's 150,000 farmers employ only one man. Many farmers cannot release them for even day-release training.

Farm bill for training scheme: £6 a worker
John Winter 29.9.67

Farmers will have to pay £6 for every full-time worker they employ in the first year of the statutory training scheme to make Britain's 300,000 farmworkers more skilful.

The levy for regular part-time workers who put in 22 to 40 hours a week will be £4. The rates are subject to approval by the Minister of Labour.

The first annual levy was announced yesterday by Mr Basil Neame, agricultural, horticultural and forestry industry training board chairman and himself a Kent farmer. He called it an "investment in our future".

The board estimates that the levy will raise £1,770,000, of which half will go in grants to farmers whose workers take training courses, one quarter on training advisory services and one quarter on administration and capital expenditure. Considering the scattered spread of farms Mr Neame considered the administrative expenses could not be considered excessive. The board had to deal with more than 100,000 employers, of whom 83 per cent had fewer than four full-time workers.

Sixteen of the board's 42 training advisers have completed their own training and are planning conferences and courses in their areas during the winter.

So far, nearly 20,000 of the 105,000 farmers who are believed to employ at least one worker have not made a return to the board. If they do not do so after getting a reminder they will receive a "notional assessment" based on Ministry of Agriculture statistics.

Mr Neame said: "We have also had several refusals to make returns. These farmers will get a reminder and in the absence of any response we shall assess their labour force and levy." Refusal to pay will compel the board to take legal action.

Farm training costs 'too high'
John Winter 13.10.67

A leading agricultural scientist who resigned from the Agricultural Training Board, criticised the board's spending on administration yesterday. Prof "Mac" Cooper, Dean of Agriculture at Newcastle University, a foundation member of the board, gave up his job in February after serving six months because he could not give it the necessary time.

On September 28 the board announced a compulsory levy on farmers for the next year of £6 for each full-time worker, and revealed that half its expected £1.7 million income will go on administration, capital expenditure and cost of training services. Prof Cooper said yesterday: "I am in sympathy with the board's aims and objectives, but not with its very high administrative costs."

A board spokesman said that since it was formed in September 1966, Prof Cooper attended only two board meetings out of six before he resigned. "How seriously therefore can we take his comments? He criticises administrative costs without allowing for the exceptional problems. In few other industries are there so many small employers so widely scattered. Once the board decided on equal opportunity for all to participate in training, collection of levies and payment of grants was bound to involve heavy expenditure from which there is no escape."

Note. Sixty farmers from many areas have signed a protest against the board's refusal to continue grants, formerly paid by the Ministry of Agriculture, to the YMCA farm training centre for boys at Egginton, Derbyshire. The centre will have to be closed next March after 35 years.

Farmers support training levy plan
Peter Bullen 20.10.67

Farmer members of the National Farmers' Union appear to have accepted the Agricultural Training Board's proposed levy of £6 and £4 a year for full- and part-time workers. But the NFU's horticultural members have inundated the union's headquarters with complaints about the levy and NFU leaders are meeting the board's employer representatives to discuss the complaints on Monday.

Mr Bill Williams, NFU president, said yesterday that his council had spent a lot of time discussing the board's proposals. So far the council had had no resolution of any note from farmers about the board. Individual council members had indicated concern at the extent of the levy, but others felt it was a worthwhile contribution for a good job which had to be done, said Mr Williams.

Letter of the week – farmers' strength of feeling
C A Harrison, Steyning, Sussex 30.10.67

You state (Farm Mail, October 20) that the majority of farmers appear to have accepted the Agricultural Training Board levy. I can assure you this is not so, and the row is only just beginning.

I will go to jail rather than pay a levy to an unelected board. My local NFU branch has just passed a resolution condemning the Board and the official NFU support for it. I have withdrawn my NFU subscription because of this. Headquarters are not yet aware of the feeling in the country: but they will be.

 # PEOPLE in the NEWS

Tom Williams, farmers' friend, dies
John Winter 31.3.67

Lord Williams of Barnburgh, architect of Britain's post-war agricultural revolution and the best-loved Minister of Agriculture in history, has died aged 79.

He had been mostly confined to his home in Doncaster since he was injured in a street accident 18 months ago. As Tom Williams, Minister from 1945 to 1951, he faced a food situation worse even than during the war, and met it with the 1947 Agriculture Act, which has been the foundation of Government support for farming ever since.

He was the tenth child of a Yorkshire miner, went into the pit at the age of 11, and continued to work as a miner and union official until he was elected MP for Don Valley in 1922. He held the seat, usually with massive majorities, until he retired from the Commons in 1959, two years before he was created a life peer.

Starting his Parliamentary career with an inbred sympathy for miners, he steadily developed an equal concern for farmers and farm workers. He was Churchill's choice as Parliamentary Secretary, Ministry of Agriculture when he formed his wartime Government.

This was when Tom developed a friendship with his opposite number in the Lords, the Duke of Norfolk, with whom he shared an absorbing interest in horse racing. The peer and the pit boy became lifelong friends.

On his elevation to the peerage he said: "I am a lord now and I live in a modest semi-detached house in the middle of Doncaster, conveniently close to the St Leger course, where I am proud to have been a steward." This was written in his autobiography *Digging for Britain* which was due for publication just after he was injured in 1965.

His close friend, Lord Netherthorpe, who, as Jim Turner, was NFU president throughout Lord Williams's career in Whitehall, said last night: "He was a shrewd politician, a fine minister, and, above all, a good friend. He won trust because he himself was trusted, and it was for good reason that he was always known as 'Honest Tom'."

Mr Bill Williams, NFU president, said last night: "Tom Williams became a legend in the countryside, and laid the foundation of the post-war development of agriculture."

The National Union of Agricultural Workers said: "He was indeed the greatest and best-loved Minister farming has ever known. The whole industry, and the farmworkers in particular, will feel a sense of personal loss."

Mr Fred Peart, the present Minister of Agriculture, said: "He did more than anyone else to ensure agriculture's well-being in the post-war world."

Laugh-a-day girl is named new Dairy Queen   *John Winter 18.7.67*

Dairymaid Christine Ginns, a 20-year-old blonde who rides a boar and a bull "for fun", became National Dairy Queen yesterday. Christine is always laughing.

"There's nothing to be miserable about," she says.

Christine, who came first among 15 finalists selected from more than 6,000 girls, was laughing yesterday as she told how she began riding Hercules, a two-year-old white boar: "One day he ran down the gangway and shot between my legs and I found myself on his back. Since then I have ridden him regularly."

She also rides Billy, an 18-month-old British Friesian bull. "Once or twice he has charged at other people, but he has never gone for me. You can get really fond of farm animals. They are very gentle, and I think they are cute."

Christine, who was crowned by the Duke of Bedford at the Guildhall, London, works for Mr Raymond Taylor, who runs a 25-acre dairy, pig and poultry farm at Unsworth, near Bury, Lancashire. She was a veterinary assistant before taking her present job a year ago. She starts work at 7 am, milks 14 cows, then tests and bottles 30 gallons of milk. Christine, who drinks one and a half pints of Jersey milk a day, also tends the calves and 12 gilts, and will look after their litters.

Christine is a member of the same young farmers' club at Bury as her predecessor, Pamela Cox, of Middleton, Lancashire. Runner-up is Miss Thelma Dyer, of King's Norton, Leicestershire, who represented the East Midlands.

Christine's prizes are £250, a Woolmark wardrobe and a three-week trip in Canada and the United States.

GENERAL

£60,000 for tickling Jack Pye's palate
John Winter 1.3.67

Millionaire builder Jack Pye and his wife prefer food grown the natural way, without chemicals. So they have just put down £60,000 to ensure the survival of the only "health food" farm of its kind in the world. New Bells Farm, 216 acres of fertile land around a 500-year-old moated manor at Haughley, Suffolk, is run by the Soil Association and is the home of "muck and magic".

Its devotees are rather proud of that phrase.

For nearly 30 years they have striven to prove that food grown their way has more flavour and health-giving elements than the great mass of chemically treated food in the national diet. In the process they have preserved a piece of land which is probably unique in this age of rationalised, artificial agriculture: 75 acres constantly farmed but, as far as anyone knows, untouched by chemicals for 30 years, and probably never.

On this farm, the manager Douglas Campbell, aged 45, and his staff produce a wide range of crops, milk, eggs and sheep. The programme is duplicated on a similar area farmed with both natural and artificial manure, while another 33 acres are kept free of livestock and receive only chemical stimulus. Cows on the "muck and magic", or organic, patch give more milk than those on the artificial fields. Latest average yields are 780 gallons, against 650 gallons. All are Guernseys, originally from the same herds.

There is a little more cream from naturally fed cows – 5.3 per cent, against 5

per cent – and they breed more regularly. Mr Campbell does not claim that all this is due to natural feeding only. There may be freakish differences in breeding, or other factors, but it calls for a scientific investigation which is beyond the scope of New Bells.

On the arable fields the "natural" yields are usually less, but there are quality differences which the association considers significant. "Natural" eggs are slightly heavier and their layers eat 1 lb less food to produce each dozen.

The farm, because of its experimental nature, runs at a loss and recently a crisis loomed. New Bells, which had been let to the Soil Association rent free, was for sale. There were no funds even to offer a commercial rent, let alone buy the farm. Then Mr and Mrs Pye stepped in.

"We had already formed a trust fund to help selected charities and we both believed in naturally produced food," said Mr Pye, 62. "We went down to New Bells and liked what we saw – and ate.

"By agreement with the Soil Association we have bought the farm and are arranging for the work to be continued and expanded. The food we have had at the farm was out of this world. It really did taste different."

With a little programming between their own organic gardens at Oxford and the vegetables, salads and other crops at New Bells they will be living on home-grown natural food all the year round.

Drugs to beat the pigeon menace
Peter Bullen 13.6.67

Farmers over a wide area of the East

Midlands are to be allowed to use a narcotic bait to trap wood pigeons next spring. Narcotic baits are designed to drug the wood pigeons which can then be collected and destroyed painlessly. Smaller birds, particularly those on the Home Office protected list, are unlikely to be affected as the bait used is normally too big for them to swallow.

For several years Ministry of Agriculture scientists have been trying to perfect a stupefying bait which will not harm other wildlife. They have conducted several limited trials but now they are ready for much wider farm testing. The wider trials are being greeted by farmers as a prelude to the general use of narcotic baits to fight wood pigeons, which cause millions of pounds' worth of damage to crops each year.

Commercial firms and local authorities have been using narcotic bait, under licence, against sparrows and feral pigeons in towns for several years. The most effective drug has been alpha-chloralose.

Letter of the week – narcotic revival? G Simmons, Frimley, Surrey 19.6.67

I assume that pigeons which take the narcotic bait in the campaign reported in Farm Mail must fall in a position where they are visible to a passing Ministry of Agriculture man.

Other unfortunate birds which take the bait will have to do the same, and also carry a small flag asking for help, and saying that they are "goodies" who wish to be revived so they can go back for a second helping.

1968

The saga of foot-and-mouth disease continued to make the headlines during the early part of the year – it did not finally go away until mid summer. Despite what Prime Minister Harold Wilson said in 1967 about the value of "the pound in your pocket", the effect of devaluation was seen in the rising prices of food. The AWB pay award was nearly caught up in the "wages freeze" but Minister Barbara Castle had to give way. Control of the use of pesticides was first mooted but Certificates of Competence were still many years away. It was one of the wettest summers on record and the "calf drain" made the headlines towards the end of what was an extremely difficult farming year.

 POLITICS and FOOD

The Prime Minister, Mr Harold Wilson, on November 19th, 1967
"From now the pound abroad is worth 14 per cent or so less in terms of other currencies. It does not mean, of course, that the pound here in Britain, in your pocket or purse or in your bank, has been devalued."

Rising prices will hit you this weekend
Peter Bullen 5.1.68

The 1968 avalanche of price increases, expected after devaluation, has begun. This weekend, supermarkets, grocery shops and stores will mark up increases on the prices of 345 items – a record for one week.

Beef prices have risen to new peaks this week and yesterday the world's second largest producer of man-made fibres, Monsanto, put up its prices by two to three per cent which will in turn put up the price of carpets, clothing and furnishing fabrics.

But the growing mass of price increases has brought the first action from the Government to prevent shoppers being exploited. Economics Minister, Mr Peter Shore, and other Government departments are watching for any hint of an unjustified price increase and any they find will be sent to the Prices and Incomes Board.

The board itself, prompted by a record number of letters of complaints from shoppers about price increases in the past few weeks, has taken steps to keep in constant touch with retail price movements. The board and the Consumer Council, an independent body financed by the Government, have both appointed liaison officers to act between them. In a few weeks the Consumer Council is starting a scheme to monitor prices in shops all over the country. The board, through collaboration with the council, will then have up-to-date information on retail price trends.

Other Government departments, particularly the Ministry of Agriculture, are also working on plans to collaborate with consumer organisations.

Increase Note: The 345 price rises in grocery shops to be announced this weekend, will affect soups, pickles, chutney, fish and meat pastes, more sweets and chocolates, beef sausages and meat pies, shoe polishes and laces, beauty preparations, ointments, bath salts, toilet rolls and kitchen paper rolls.

A victory for the pure food lovers
John Winter 31.1.68

Pure food enthusiasts, who dislike any fruit or vegetables produced with artificial fertilisers or chemical pesticides, made a major breakthrough yesterday. One of Covent Garden's leading firms announced that it had agreed to set up a section of the market for organically produced fruit and vegetables.

The Soil Association, which arranged the discussions that led to the firm's decision, said that produce compost grown without sprays, which hitherto had been available in only a limited number of shops, would now be on sale to a very much wider public. Mr Michael Allaby, the association's spokesman, hopes it will eventually lead to the sale of organically grown food in greengrocers' and supermarkets all over Britain.

The association hoped more farmers and growers would devote at least part of their land to growing fruit and vegetables organically.

Farmers fear Whitehall land grab
John Winter 21.2.68

Hundreds of farmers fear they could be victims of a Whitehall land-grab. Some have formed an organisation called The Freedom Fighters.

The rumpus has blown up in Mid-Wales where the proposed Rural Development Board may be called to administer some 1,300 square miles in five counties. The board plans to revitalise farming and forestry with voluntary schemes to merge farms into larger units. But the farmers think that the board could use compulsory powers to buy out the "small man".

Vice-chairman of The Freedom Fighters, Mr Joseph Wright, 55, said yesterday that one form of protest being considered would be refusal to complete statutory returns.

The Farmers' Union of Wales will represent 1,700 objecting farmers at an Aberystwyth public inquiry in April to consider the board's boundaries. Mr Gwilym Thomas, the FUW spokesman, said at Aberystwyth: "Farmers see this as land nationalisation. I don't know of any subject which has caused them so much feeling."

Call charges
John Winter 21.2.68

Packing station owner Mr Colin Ritson, of Brampton, Cumberland, says the Egg Marketing Board office in Leeds made a 3s 6d telephone call to tell him one egg was missing from his weekly return of about 216,000 eggs. Value of the egg: 1½d.

No prizes for Peart
John Winter 7.3.68

Britain's farmers took a long, critical look at the farming price review and its unprecedented £52.5 million carrot yesterday. Then they sent the Agriculture Minister, Mr Fred Peart, their deflating verdict. It read: "The National Farmers' Union regrets that this will not enable the industry to play the part it could do in the present economic circumstances."

In a statement the council of the NFU said: "An opportunity has been missed to make doubly sure that devaluation succeeds, by stepping up the agricultural import-saving programme. For foods which can be grown in this country, the effect of devaluation may ultimately be to increase food imports by about £100 million."

The farmers say the award falls short by £16 million of the figure to cover the increased costs of the past year. Mainly through devaluation, costs zoomed £68.5 million to their highest-ever total on review products. Biggest items on the bill were £17 million on labour, £14 million on feeding stuffs and £11.5 million on fertilisers.

Yesterday's NFU verdict was reached after a four-hour wrangle by the 200-strong "jury". For the first time in 21 years they declined either to agree or disagree with the annual settlement. Previously they agreed with 14 reviews, while six were said to have been "imposed".

The union hoped that new prices would be fixed giving strong incentives to home food production, to replace £200 million in food imports. But the Government feels the settlement is as generous as farmers could expect in the economic situation.

As I predicted, the NFU decision was touch and go. Farmers, already soured by the decision to resume imports of disease-suspect beef, were inclined to look coldly at this collection of sops for everyone, which was not substantial enough to please anyone.

But the Ministry feels that the farmers are in fact £14 million in pocket with this review. It maintains that to refund all cost increases would be unrealistic and that farmers must cover the £16 million by increased efficiency estimated at £30 million a year. The review says farm incomes have risen £18.5 million to £510 million, and that total output went up by six per cent in 1967–68 after a three-year standstill.

Farming productivity which slowed to a four per cent rise last year is up this year by a splendid seven per cent.

This is how the review's cake was sliced. The biggest piece goes to beef – 11s per cwt up on the end price, plus an extra 30s subsidy on beef calves, and £2 on hill cows. The link with milk, bound to go up if beef moves ahead, is recognised by a guaranteed rise, plus standard quantity adjustment of 1¼d a gallon. This will be passed on in the form of a ½d pint increase in retail prices from July.

Sheep farmers get a rise for meat, but a standstill on wool, which had a bad year. Pigs, now expanding fast, get increases enabling farmers to produce 300,000 more without penalty.

Handsome rises for cereals, especially wheat, will stimulate home production. And the standard quantity limiting guarantee commitment is abolished. Potatoes get an unexpected fillip. If 1968 shows a crop surplus the Government will put another £1 million into the support scheme.

Only the poor old egg farmer takes a price cut of about a halfpenny a dozen – because over-production is expected.

Milk levy   *John Winter 7.6.68*

Sir Richard Trehane, chairman of the Milk Marketing Board, yesterday defended the board's proposal to increase the maximum capital levy on milk producers from a farthing to a halfpenny a gallon. He said the extra money was needed to modernise existing creameries and build new ones.

Milk sales slump   *John Winter 18.10.68*

Britain's cows produced a record 174 million gallons of milk last month, but sales slumped by 22 million pints, the Milk Marketing Board said yesterday. It blamed price rises, the decision to stop free milk in secondary schools and a late start to school terms. This is the fourth month running that sales have dropped.

CROPS and MACHINERY

Farm crops and soil gone with the wind   *John Winter 2.4.68*

Farmers whose top soil and 1968 crops have been swept away by strong winds are appealing for help. Mr Fred Peart, Minister of Agriculture, is considering their plea.

The worst soil blow for 25 years stripped thousands of acres of fen and light land in Lincolnshire, Holland, Norfolk and Isle of Ely of their top soil, seeds and seedling cereal plants and deposited them in dykes and ditches. Here the precious soil has created a threat of flooding at the first heavy rain, and farmers are working feverishly to reopen the drainage system in time.

Mr Bill Williams, National Farmers' Union president, has asked Mr Peart if they can have a 50 per cent Government grant towards the cost of the work. Normally such payments can be claimed only for improvement work which is approved in advance. In the present emergency Mr Williams asks for retrospective grant aid, "bearing in mind the paramount importance of immediate remedial action."

Mr Maurice Brewster, NFU assistant secretary in Lincolnshire, where thousands of acres of top soil blew away, said: "Spring work was never so well advanced. Frost and dry weather had given an excellent seedbed. Seeds had been drilled and the early ones were through. Then whole areas were stripped and dykes 5ft deep were filled to the top with a mixture of soil, seeds and fertiliser."

The cost of clearing such a dyke is £5 for 22 yards. Some big farms have as much as 2,200 yards to be cleared.

Greens are caught in cold blast

John Winter 15.4.68

The cold spring is playing havoc with early vegetables and salad crops.

Unless it gives way quickly to warm, moist, growing weather shoppers will face high prices for lettuce, radishes, cabbage and cauliflowers, whose growth is at a standstill. Early potatoes, too, are falling behind, though there is time for them to catch up to an average harvesting timetable.

East winds, sharp night frosts and lack of rain have persisted almost without a break since mid-February in the early growing areas of Worcestershire and in the West Country. Mr Sidney Mence, chairman of Pershore Growers, the Market Gardeners' Co-operative, said yesterday that he had never known a spring cold spell of such severity to last so long.

In the Vale of Evesham market gardens it has killed off many early planted greens and growers are using irrigation to get crops into the ground and keep them alive. Outdoor lettuce is bound to be late and a gap could occur between the end of the glasshouse crop and the first field crops. Mr Mence said: "Outdoor radishes look even worse. Last year at this date we were pulling them. This year they will not be ready for another fortnight and then there will be only half a crop. Brussel sprout plants are very bad. Some of the cabbage started under glass and planted out has been killed and early cauliflowers look terrible."

The weather is already affecting prices. Spring cabbage, which at this date last year was making 5s a crate in Pershore Growers' market, is from 10s to 14s, though not of such good quality.

Phantom tractor takes to the fields   *John Winter 22.4.68*

A phantom tractor which can work without a driver on ploughing, harvesting and other field operations is expected to start trials later this year. It is the brain-child of researchers of the National Institute of Agricultural Engineering at Silsoe, Bedfordshire, who are working on automation techniques which would release manpower for other tasks.

They have solved the problem of guiding the tractor on a course parallel to the profile of whatever work it is doing, such as the wall of a ploughed furrow or the edge of a standing crop which is being harvested. This is done by an instrument on the machine which receives an ultrasonic pulse echo from the profile.

The problem that remained was to "tell" the tractor how to slow down and turn round when it reached the edge of the field. The answer is an optically reflective "fence" along the field perimeter and two special lamps on the tractor. These pick up reflections from the "fence" as the tractor approaches, which are translated to the controls.

Spray Eagle swoops off with cup   *John Winter 28.5.68*

Mr Neil Harvey, 31-year-old farmer and agricultural contractor, took his new machine Spray Eagle with a 60-foot "wing span" to the Bath and West Show at Shepton Mallet, Somerset, yesterday and nearly swept two judges off their feet.

Spray Eagle looks like one of those old tank fire engines, except that fixed at the front is the longest spray boom the judges have ever seen – 60 feet of it. This is the width of the strip of land which Mr Harvey can cover with weed killer or other chemicals.

When the judges, Mr F H Garner, principal of the Royal Agricultural College, and Mr T Sherwen, president of the Institution of Agricultural Engineers, learned that the machine had trebled his work rate they had no hesitation in placing it first of 14 entries in the show's new ideas competition. This means that when the show opens tomorrow Mr Harvey's firm, Michael Pritchett, Meadowlands Farm, Bibury, Gloucestershire, will receive the *Daily Mail* Life Challenge Cup and a cash award of £100.

Spray Eagle has already sprayed 2,000 acres this year at a rate of 30 acres an hour. The operator sits in a lorry-type driving cab, completely protected from weather or drifting spray and has a clear view of all 40 nozzles on the boom except five in the centre.

Second prize was shared between Edwin Parrott, 20, of Glanville Wootton, Dorset, for his automatic cow-shed cleaner and W N Walford, Long Sutton, Langport, for an automatic mobile feeder of milk substitute for individual penned calves.

All the inventions will be on show throughout the four-day show.

Flowers get a robot 'nurse'

Peter Bullen 1.8.68

The robot horticulturalist is here – a small, £500 computer which nurses flowers or vegetables night and day. The computer, Solplan Mark III, is programmed to provide the heat, air, gas and water a greenhouse crop needs in varying conditions.

It automatically switches on the heaters if the plants are cold, or opens the ventilators if they need cooling. It controls watering and humidity and supplies carbon dioxide, which enriches the air and boosts growth. A warning device still in the design stage would warn the glasshouse owner if the computer could not cope with a situation.

Computers are already in use at the Ministry of Agriculture horticultural research station at Lymington, Hampshire, and in one or two leading growers' glasshouses. The manufacturers claim that they will increase crop yields and quality and cut costs – and the crops will have as much flavour as any. Installed in a third of an acre of glass, a computer could pay for itself in a year.

Solplan managing director Mr A C Craig says that no other country has developed such a fully integrated control system for greenhouse environment, and there is a big export potential.

The trickiest corn harvest for years

John Winter 13.8.68

Thousands of combines were driven into cornfields yesterday in an all-out attack on the trickiest harvest for years. Heavy summer rain, hail-storms and floods have robbed many farmers of what promised to be heavy yields of good quality grain. In the South and East a wet, sunless fortnight has encouraged disease and growth of weeds which have reduced fields of flattened corn to an appalling tangle. But in other areas, chiefly the West Midlands and North, there are many promising fields of undamaged grain.

Farmers are now putting their machines to work immediately the soft ground is dry enough to bear them. Any further prolonged rain could mean a write-off for many crops. In Devon Mr Murray French, Exeter corn merchant, said: "The harvest is going at fairly high pressure. Barley from fields which went down is not so good, but there are some nice crops from fields which escaped."

In Suffolk farmers are still working on winter barley, which is normally completed by this date. Yields vary from 30 cwt to 40 cwt an acre, but the biggest anxiety is that signs of sprouting, due to excessive humidity, are already showing in the ears.

In coastal areas of the South, which normally have the earliest harvest in Britain, barley crops have had the most damage, and yields are down. Wheat, on which farmers have usually made a good start by now, is still untouched, but looks well. Oats are excellent.

Frustrated farmers dry out and wait for wheat   *John Winter 14.8.68*

Gloom spread across harvest fields yesterday as rain returned and brought combining, already a fortnight behind normal, to a dead stop. And sharp heavy rain storms took a further toll of yield and quality from the ripe crops.

The only sign of progress was the drone of drying machines in farm buildings, working continuously to draw surplus moisture from the few loads of grain so far gathered. Frustrated farmers, already reconciled to reduced yields from barley crops, are now pinning their hopes on remarkably good looking fields of wheat, if they get the weather to harvest them.

Oats, which are only 6.5 per cent of the national cereal acreage, are giving excellent yields to farmers who have grown them. Field beans, which jumped 67 per cent with the inducement of a £5-an-acre Government subsidy, have, in Hampshire, been seriously hit by chocolate spot disease.

Mr John Loader, sales director of James Duke and Son, Bishops Waltham corn merchants, said his firm had so far handled nothing but winter barley, with a few hundred tons of spring barley and oats, and their first crop of wheat probably about 5 per cent of total intake.

Mr Tom Hewer, with 1,000 acres of corn, more than 90 per cent of it barley, at Chilcombe Manor, near Winchester, has so far harvested only 120 acres. Last year he had combined the lot by the end of August.

The sun brightens the harvest hopes   *John Winter 24.8.68*

Farmers throughout the country have planned the biggest harvest blitz in history for this weekend. Alerted by a forecast of continuing fine weather, they will be working flat out to gather weather-damaged corn crops and catch up with a harvesting timetable which is so far two weeks late.

The heat wave, which has now lasted three days in most areas, has already transformed the picture. More corn has been cut than in the preceding three weeks and gloomy forecasts of a harvest disaster have receded. Estimates of progress range from a cautious 10 per cent overall by the Ministry of Agriculture to double that figure by NFU regional officers in the South-West and North, and 15 per cent in the granary of East Anglia.

Yields and quality of grain are still expected to be only average or worse, but many farmers have been agreeably surprised at the harvest they are getting from fields which have been flat for weeks. Mr Peter Savory, NFU cereals committee chairman, last night had completed half the 500 acres of cereals on his farm at Saxlingham, Norfolk.

He said: "I am probably further ahead than most, because I doubled my combine strength to two 12ft machines this year. Three more fine days will see us through the barley, leaving 145 acres of wheat. Yields have varied tremendously from a very bad 22 cwt an acre from one field of spring barley up to 36 cwt on another. Progress is slower than usual because crops have been so badly battered, but this wonderful weather has changed what could have been a disaster to something a good deal better."

In the South-West progress ranges from 15 to 30 per cent in different areas and in the North from 8 per cent on the Scottish border to 20 per cent in Cheshire.

The scarecrow walks round the field
Peter Bullen 4.9.68

Britain's first mobile scarecrow was introduced to farming yesterday. From a distance he looked like a jockey in scarlet colours riding a tricycle backwards. Closer inspection showed that he was a dummy bolted to three wheels powered by a small lawnmower motor. He is called "the Warden", and travels all day long in a large, pre-set circle round and round the field, scaring the birds wherever they settle in a way that would put the traditionally rooted scarecrow to shame.

This scarecrow is the brain-child of one of Britain's most prolific inventor-farmers, 61-year-old Mr Roy Rothery, of Walcot, Lincolnshire. Like all his inventions, the scarecrow arose from a problem Mr Rothery faced on his own 350-acre arable farm. Last winter he lost £1,500 worth of brussels sprouts to wood pigeons.

Mr Rothery said birds soon became accustomed to static scarecrows so he developed one that moved. A refinement would be an adjustment to make the exhaust noises louder to help to scare the birds. He estimated that it would cost £75 to make but would soon pay for itself in extra crop yields.

"The Warden" was only one of five inventions that Mr Rothery unveiled yesterday. The most impressive was a device to deliver seed grain straight from a trailer into the seed drill at the touch of a switch.

Mr Rothery said that until now seed had always been humped in sacks for loading into the drills. He said: "I call that serfdom. There are two million seed drills in Europe, all needing a serf to load them by humping large sacks. This device protects the worker and can fill up a drill with 4 cwt of seed or fertiliser in a couple of minutes."

The cost of the device to the farmer with his own motor and auger would be about £150.

Mr Rothery, who has already had about 15 ideas marketed, including grain cleaners and driers which have been bought by thousands of farmers in Britain and overseas, also demonstrated a new rice cleaner, sugar beet precision spray, and an automatic grain-loading device.

Tractors will save lives and cut deafness
John Winter 4.10.68

Farm tractors which will not only save drivers' lives if they turn over, but also protect their hearing, are on the way.

Fitting of safety frames or cabs to all new tractors becomes compulsory in 1970. The Ministry of Agriculture has decided this is the only way to protect drivers from this kind of accident, which has already claimed 22 lives this year. Manufacturers' ideas to meet the new rule will be seen at the Royal Smithfield Show, which opens in London on December 2, and some safety cabs will incorporate ideas to reduce noise.

Research at Loughborough University has proved that half the tractor drivers in a survey suffered from deafness. A driver sitting in an enclosed cab is exposed to even more noise than on an open tractor. The National Institute of Agricultural Engineering has not so far tested a safety cab which reduces noise below the level of an open tractor. But it has produced a test-bench model which cuts noise to about half this level.

LIVESTOCK and POULTRY

Farm disease valuers get sack

John Winter 3.1.68

Several valuers have been barred from assessing animals slaughtered in the foot-and-mouth epidemic. They have been taken off the Ministry of Agriculture's list of valuers.

Values have increased by up to 50 per cent since the outbreak started on October 25. The Government is now due to pay farmers more than £20 million compensation. At least one of the valuers has protested to the Ministry and to his own association, the Incorporated Society of Auctioneers and Landlord Property Agents.

The Ministry is re-examining reports on the 2,200 cases in the epidemic to determine when and why the rising valuations started. They date from about mid-November. The Ministry also faces complaints from farmers who were early victims of the outbreaks. Although satisfied with their compensation at the time, they are now demanding it should be adjusted to present levels. The NFU is supporting them.

While the Ministry accepts a rise in values as slaughterings mount and the demand for replacement grows, its veterinary officers in the field consider that some recent valuations have gone beyond all reason. Vets in charge of outbreaks, usually the senior Ministry official on the spot have, by and large, accepted valuers' figures. But certain recent cases have been referred to a senior regional veterinary officer, and, in some instances, a new valuer from a different firm has been called in.

Mr Smith Wright, deputy secretary of the Incorporated Society of Auctioneers and Land Agents, said one member had reported that he was threatened with removal from the Ministry list and had protested. He said: "Valuation in the circumstances prevailing is a very difficult business. I suppose the only fair way would be to have a panel of three or four valuers, but this would be quite impossible in an epidemic like this. Some rise in value is obviously justified but I would not have thought it should have reached the reported level of 40 per cent."

One suspended valuer said: "My valuations have risen about 20 per cent over the period of the epidemic. This is based purely on my professional opinion, having regard to the scarcity value which is developing."

● For the third day in succession 11 new foot-and-mouth cases were reported in 24 hours up to 5 pm last night. The total cases reached 2,198, and the animals slaughtered are 179,540 cattle, 80,113 sheep and 97,056 pigs. Shipment of Irish beef cattle, both fat animals for immediate slaughter, and stores for rearing, will be resumed through Birkenhead and Heysham from midnight tomorrow.

Letter of the week – congratulations

Mrs Suzanne Herring, Somerton, Somerset 3.1.68

I really should like to congratulate you on the tremendous way in which you have made the general public aware of this awful foot-and-mouth disease disaster without being in the least bit maudlin.

As a farmer, I cannot thank you enough for this great piece of journalism. I have asked readers of other papers if their coverage has been as great, and it obviously has not.

In particular, I was very thrilled over Mr Walter Bromfield's idea for helping farmers to re-stock. It must be heart-breaking to see 20 or more years of work destroyed.

The Queen helps to restock herds
John Winter 11.1.68

The Queen has offered to help restock farms stricken by foot-and-mouth disease. Her offer was made yesterday in a message to Mr Fred Peart, Minister of Agriculture.

The Queen said she was greatly concerned by the severity of the epidemic. Her message went on: "I understand that there is a scheme for those farmers who, at the end of this present epidemic find themselves unaffected, to be able to help those who have lost all their animals. I wholeheartedly support this effort and when the time is right I wish to contribute to this pool of livestock from the royal farms. I am sure that this scheme will appeal to a very large number of farmers."

No animals from the royal farms at Windsor, Sandringham and Balmoral have yet been earmarked for the restocking scheme, but a score or so are likely to be offered. They will be put on the register of stock available for purchase at reasonable prices by any of the stricken farmers.

At Windsor there is a herd of 126 pedigree Jerseys and 73 Ayrshires, 30 Large White breeding sows and a flock of 300 breeding ewes. At Sandringham there is a large herd of commercial beef cattle, and at Balmoral a herd of pedigree Highland cattle and a small herd of the new Luing beef breed.

The Queen is unlikely to make a gift of animals from the royal farms because of the difficulty it would cause in selecting recipients from among the 2,250 and more farmers. Mr Peart thanked the Queen for her message and said the Ministry hoped to make a start on restocking soon.

Nine new cases of the disease were confirmed in the 24 hours up to 5 pm yesterday, including one in Warwickshire – cleared as an infected area from the end of December. Thirty-nine cattle and 23 pigs were destroyed on the farm at Brandon, Coventry. A further case was reported after 5 pm in Derbyshire, an already infected area.

Last night, the Eire Minister of Agriculture, Mr Neil Blaney, warned Irish Rugby fans travelling to the Twickenham game on February 10 that on their return they would face the 21-day movement restriction order affecting travellers from Britain.

Doubly unlucky
John Winter 15.1.68

Farmer Mr Sidney Simpson, of Llynclys, near Oswestry, who lost 100 pedigree Friesian cattle in the foot-and-mouth epidemic, suffered a second blow at the weekend when 1,000 of his turkeys were destroyed in a farm blaze.

Hard-hit farmers plan to oust Egg Board
John Winter 13.2.68

A campaign to kill off the troubled Egg Marketing Board even before the Government inquiry into its future is completed, has been started by hard-pressed egg farmers.

Forms demanding a poll of the Board's

registered producers to decide whether the Board should be wound up are pouring into the National Egg Producer Retailers' Association. One thousand signatures from the 246,000 producers on the register are needed to secure the poll, but the demand will not go to the Board until the Association has made a final appeal to Mr Fred Peart, Minister of Agriculture, to save producers from ruin.

A letter has gone to him outlining the desperate situation they face following a cut last Friday of 3*d* to 4*d* a dozen, according to size, in the prices the Board pays for their eggs. Mr John Furniss, Association secretary, said: "This would bring disaster to thousands of egg producers. They just cannot stand another year as bad as the one just ending. For nine of the 12 months, prices have been so bad that farmers have had to sell eggs for less than the cost of producing them. After last week's cut it looks as though the price pattern next year will be worse. We contend this is due to gross mishandling of the market by the Board.

"We have appealed to the Minister to discuss the situation with us. If he cannot give us hope of early improvement, I fear our members – and we have more than 2,000 with ten million laying birds – will decide to go straight ahead with the demand for a poll. We are nearly home and dry with the necessary signatures already."

Yes it was Argentine meat

John Winter 28.2.68

The foot-and-mouth epidemic, which cost Britain £150 million and led to the slaughter of 422,300 animals, *was* caused by Argentine meat. That is the finding of a team of Ministry of Agriculture experts after an investigation lasting several months. Last night Mr Fred Peart, the Minister, was studying their report before giving details to the Cabinet tomorrow.

Although most blame is levelled at lamb, Argentine beef does not escape criticism. Because of this vets want to see a total ban on all meat from suspect countries continued. But they believe Mr Peart will have to compromise by banning Argentine lamb and allowing the import of beef. The three-month ban on imports, introduced on December 4, runs out on Monday.

The report shows that Argentine beef and lamb were widely distributed in the disease areas in the crucial two to three weeks before the epidemic started. The plague did not start with a single outbreak but from several primary outbreaks widely separated.

Veterinary investigators established that Argentine beef had been delivered to farms struck by the disease. They also found that Argentine lamb was being distributed after being boned and that bones ending on rubbish dumps were causing infection. The Argentine Government has a vaccination scheme for part of its cattle population, but not for sheep. Last year Britain imported 18,000 tons of Argentine lamb, 3 per cent of our supplies, worth £3 million.

There were no new outbreaks of foot-and-mouth in the 24 hours to 5 pm yesterday, the sixth successive clear day.

The big move to restock farms gets under way
John Winter 29.2.68

Hundreds of cattle will this weekend hit the great restock trail on the farms which have been empty since the foot-and-mouth epidemic started more than four months ago. Yesterday Ministry of

Agriculture disinfection squads moved into farmyards near Oswestry, Shropshire, to spray hayricks and other feeding stuffs with Formalin.

The go-ahead signal reached Oswestry just in time to avert a sitdown protest by farmers in the local disease control centre. They were angry over the hold-up on restocking licences which followed the recurrence of disease on three farms in other counties which also had it before Christmas. The decision on whether to lift the ban on imported meat from "suspect" countries will be announced on Monday. MPs were told this yesterday by Mr Fred Peart, Minister of Agriculture.

No fresh cases of foot-and-mouth disease were reported in the 24 hours ending at 5 pm yesterday, the seventh successive day without an outbreak.

The sun shines again on that farm where 12 weeks ago the only animal alive was Nell the collie
John Winter 9.3.68

A family whose world collapsed just 12 weeks ago were looking confidently forward to the future yesterday. And they saw it in the shape of one frisky calf.

Its arrival marks the end of the long, sad silence which fell on Prospect Farm, Nantmawr, Shropshire, when foot-and-mouth disease wiped out the Sevenwells herd of pedigree British Friesians. Mr Harold Jones, 43, had devoted most of his farming life to that herd. When it was destroyed he and his wife, Sylvia, 35, lost their entire income and the 12-hours-a-day, seven-days-a-week job that they loved. The only animal left alive on Prospect Farm was Nell the collie.

They were stunned with grief when I first met them last November. But the picture which Chris Barham took that day brought the first hint of a new life which would come to the 50-acre farm. The picture touched the heart of Mr Walter Bromfield, 58, who has 500 British Friesians on two farms near Langport, Somerset. He wrote immediately to the Joneses saying: "Let me help you to make a fresh start."

Mr Bromfield gave Mr Jones the £200 bull calf which is now at Prospect Farm. It comes from one of the finest cows in his Earnshill herd, which holds the Friesian Society's highest milk qualifications. The calf, now six weeks old, will grow into the bull that Mr Jones hopes will start the new Sevenwells herd. As well as the calf, Mr Jones also bought six cows from Langport to add to the six heifers which he had ordered from another herd.

Mr and Mrs Jones with their children Jennifer, seven, and Alan, five, have since been on holiday in Somerset and spent two days on Mr Bromfield's farm. Mrs Jones said: "I think Mr Bromfield is the kindest man I have ever met."

Mr Bromfield said: "I believe we have undoubtedly made a lifelong friendship."

The name of the calf? By common assent it has been registered as Earnshill Daily Mail.

'More milk' flies in
John Winter 6.4.68

Golden Guernsey cattle, introduced to Canada from Cornwall 80 years ago, came back yesterday in test tubes.

The English Guernsey Society, fighting for a bigger share of the national dairy herd, flew in semen of two of Canada's top bulls to raise the milk production and increase the size of its animals in Britain. Farmers will pay £5 a time to have their cows inseminated, and

160 of the 1,125 shots on the jet were sold before it touched down at Heathrow. The bulls are expected to sire daughters, which will yield an extra 70 gallons of milk with each calf.

In foot-and-mouth-stricken Cheshire five new Guernsey herds have already been established. Eight more will be founded from 100 cattle to be shipped from Guernsey.

Livestock food men warned on prices
John Winter 1.5.68

Producers of food for livestock were warned not to raise their prices unjustifiably last night by Mr Cledwyn Hughes in his first speech as Minister of Agriculture. Speaking at the biennial dinner of the Compound Animal Feedingstuffs Association in London, he agreed that the industry faced increased prices for imported raw materials. "I do not propose, ostrich-like, to hide my head in the sand and ignore commercial facts. Some prices will, of course, rise, not only because of devaluation, but also as a result of some measures in the Budget and for other reasons.

"It is quite essential that prices should not rise unjustifiably. This is not a political point. Whether in government or industry, we cannot escape the conclusion that this is of great importance to the future economic health of our country."

He urged the feed-makers to use more home-grown wheat and barley.

Farmers' chief refuses to quit
John Winter 4.5.68

Mr Henry Plumb, deputy president of the National Farmers' Union, has refused to resign from the Northumberland Committee inquiring into the foot-and-mouth epidemic. Butchers complained when he advised housewives to boycott meat from countries where the disease is endemic. Mr Plumb also refused to stop making further comment on this explosive issue while the committee were sitting.

Mr Ken Forder, secretary of the National Federation of Retail Meat Traders, asked Mr Cledwyn Hughes, Minister of Agriculture, to relieve Mr Plumb of his committee place. He was appointed, with other members, by the former Minister, Mr Fred Peart.

Mr Plumb, who is at present acting leader of the NFU, said: "I represent farmers. Our views on the risk of importing disease in meat are well known. It would be quite impossible for me to seal my lips on this issue while the committee is at work, which may well be for a considerable time. I will certainly not resign."

The NFU foot-and-mouth action committee is co-ordinating the efforts of county branches to strengthen the boycott call.

Birthday treat – Albert's monster pig
John Winter 18.5.68

It was Mr Albert Clements's 40th birthday yesterday and it was marked by his 1,000lb 8ft-long boar being judged the best of 243 pigs at the Devon County Show at Exeter. The 2½-year-old Large White, Shimpling Field Marshall 18th, won the Exeter Agricultural Society's cup.

Neither the boar nor Mr Clements, who runs a 70-acre family farm with 50 breeding sows at Bardwell, Suffolk, had been to Devon before. The judge for the supreme title, Mr Bill Whidden, could hardly believe his eyes as Field Marshall made his entry. He said: "We have not had such a magnificent Large White boar in the West Country for as long as I can remember."

With a string of seven pigs this completed for Mr Clements a tally of three championships, four firsts, one second and one third prize. All the animals are from a new herd he started after his old one was slaughtered in the swine fever eradication scheme in 1963.

Success also came to Mr Andrew Meikle, 41, with the first animal he has ever entered at the show, his six-year-old Ayrshire cow, Meikle's Maisie. She was judged the best dairy animal of any breed. She is the pride of the 100-strong herd Mr Meikle and his wife founded eight years ago on a 103-acre farm at South Molton, in which they invested their £2,000 savings.

Royal Jerseys for Miss Doreen

Peter Bullen 10.6.68

Five cows from the Queen's prize-winning Jersey herd, reared in the shadow of Windsor Castle, were grazing on a 40-acre Flintshire farm yesterday. Their new owner: Miss Doreen Lowther, of The Grange, Overton, who had her own herd of 50 Jerseys slaughtered in the first few days of the foot-and-mouth disease epidemic early last November. For almost five months Miss Lowther waited on her silent farm for the epidemic to die down and for the day restocking could begin.

And early in the epidemic that destroyed 430,000 animals on 2,365 farms and almost as many neighbouring "contact" farms, the Queen said that she would like to help some of the victims. When the National Farmers' Union announced its list of farmers willing to sell animals at reasonable prices to help the victims, 35 beef and dairy cattle from the royal herds were immediately made available.

Pick of the 35 were ten young heifers in calf from the Windsor herd which Queen Victoria founded 100 years ago. As soon as it was safe for Miss Lowther to visit the Royal herd she went to Windsor and selected five. Neither she nor the royal farms will disclose the price but it was certainly not more than half the £600 the five young cows are worth.

Recently, the five were delivered to The Grange where the first four cows Miss Lowther had bought from other farmers have already started producing rich, creamy milk for which the Jersey breed is famed. Her own Iscoyd herd used to produce an average of more than 820 gallons of milk each year – the highest output of any recorded Jersey herd in Flintshire or Denbighshire. The milk contained more than 5 per cent butter fat and over the years Miss Lowther built up a thriving farm gate sales trade.

Egg Board faces delayed death sentence

John Winter 18.6.68

Disappearance of the Little Lion and the end of cheap eggs for many housewives are likely to follow the publication on Friday of the report of a Government commission on egg marketing. The report is expected to recommend the winding-up of the trouble-torn Egg Marketing Board and the setting up of a central authority whose main job would be to buy eggs at times of surplus production.

Mr Cledwyn Hughes, Minister of Agriculture, has given organisations until July 31 to give him their views on the report, but the feeling that he will accept its main recommendations is so strong that a situation of disorder could develop before the new set-up comes into being.

Thousands of farmers are already breaking the rules about selling un-stamped eggs, because this way they can get better prices than if they sell to the board. This has caused trouble for some of the 250 packing station companies, which are the board's agents. The board, faced with a delayed death sentence lasting until at least next April, is unlikely to be able to enforce its regulations on packers any more successfully than it has on producers.

Mr Vallence Collins, National Egg Packers' Association chairman, said: "Anyone who is not already making plans is very shortsighted. I think there is a real danger if the recommendations which are rumoured are not implemented very speedily."

If the Egg Board goes, the biggest losers will be egg farmers in areas remote from main consuming centres and the smaller packing companies.

Was it sabotage?   Peter Bullen 29.6.68

Britain's worst epidemic of foot-and-mouth disease could have been spread intentionally. The possibility has been raised in the British Veterinary Association's evidence to the Government committee investigating the epidemic.

The BVA said: "An unvaccinated national herd is extremely vulnerable to deliberate introduction of virus by an enemy in wartime or by subversive elements in times of international unrest. Even in the recent epidemic there were rumours of some malicious spreading of the infection, but it is impossible to know whether they had any foundation in fact."

A number of people had suggested that a unique feature of the epidemic was that some unknown factor was operating at the outset which led to early, rapid dispersal of infection, said the BVA. The BVA's own theory is that the only significant unusual factor was that the site of the primary outbreak and the prevailing weather conditions carried the infection into a heavily stocked area.

The Little Lion men start fight to survive   Peter Bullen 9.7.68

The Little Lion men who form the Egg Marketing Board started to fight back yesterday against the report of the Government reorganisation commission that the board should be wound up. The country's 194,000 egg producers registered with the board received a leaflet from its chairman, Mr Geoffrey Kidner, urging them to lobby Press, Parliament and farming organisations to save the board.

Mr Kidner said: "Remember that the industry has only to the end of July to make its comments to the Minister. There is not a moment to lose."

The board's chief information officer, Mr David Smith, told a meeting of producers in Kirkwall, Orkney, last night: "Make no mistake about it. If the report is implemented in its present form, a lot of people who are today making a reasonable living out of eggs will be looking for other jobs tomorrow."

If producers scrapped their elected board and put in its place an appointed authority suggested by the report they would be handing over the industry to the big boys and the trade to carve up between them.

Mr Smith, who urged the Little Lion producers to roar their disapproval, said: "The board is resolved to fight on these issues. It is not our wish to see the bulk of our producers thrown to the wolves.

But we cannot fight alone. If you share our concern about the implementation of this report, I hope you will raise such a roar that Whitehall cannot fail to listen."

Fighting porkers get bridal suites
Peter Bullen 25.7.68

A "honeymoon hotel" for pigs where each sow has her own room complete with running water in which she can entertain a male visitor was shown to hundreds of pig farmers yesterday.

The building is at Whitethorn Farm, near Stoke Mandeville, Buckinghamshire, scene of the first National Pig Fair. About half the 6,000 visiting farmers toured the farm and saw over the building.

The idea was to keep sows snug and comfortable and above all by putting them in individual pens to prevent them fighting. The hotel holds 30 sows and two boars who are allowed out for a few minutes each day to walk the central corridor and to visit any of the sows of their choice.

Mr Sandy Barker, farm manager for British Oil and Cake Mills, the fair's organiser, said: "You must give sows individual attention at mating times." He has been running the "hotel" for six months. The sows spend a month feeding and resting in the hotel and already there has been noticeable improvements in their condition and they are producing larger litters. Mr Barker said a full circle had been completed which had begun with looking after individual animals; then large number of animals and now back to individuals.

The National Pig Fair at which 100 firms exhibited and with 6,000 visitors looks like being an annual event in the agricultural calendar.

Letter of the week – baloney!
W York (Full name and address supplied) 19.8.68

I am fed up with all this blah about fresh eggs, brown eggs and free-range eggs and with criticism of the poor maligned lion egg. I will challenge any independent tasters to distinguish one from the other in an impartial test of eggs I will provide. This goes for eggs in their shells (with the lion stamp turned out of sight), or fried, boiled, poached, scrambled or cooked in any other way.

Lion eggs are good eggs, and all this baloney is just a trick to increase the price of unstamped eggs to a gullible public.

Government puts clamp on massive calf drain
John Winter 14.10.68

The Government has decided to crack down on the calf drain to the Continent. Sometimes more than 3,000 animals are being shipped abroad each week. The Ministry of Agriculture is working out a way to greatly reduce this and an announcement is expected this week.

The animals are going to Common Market countries where they are raised for veal or beef. The trade, which has brought in more than £500,000 in foreign currency since it started in August, has priced good beef calves above the economic level which our own beef farmers can pay, and threatens to push up home beef prices in 10–15 months' time.

It is also said by the RSPCA to flout existing regulations and Government proposals to safeguard calves' welfare, and to provide European farmers with animals for rearing into "white veal", on a system condemned in Britain. This claim is based on reports that many calves are going to countries which specialise in white veal production by feeding them on an unbalanced diet deficient in iron.

The Ministry of Agriculture is to limit the trade by raising the minimum weight for export calves. This would make them unsuitable for white veal production.

Control of exports is operated by two departments, the Ministry and the Board of Trade. The Board, which issues export licences, is concerned only with the economic side of the operation. But the Ministry of Agriculture, to which exporters must apply for veterinary clearance of each calf consignment, has a detailed set of welfare regulations which are implemented by a Ministry veterinary officer, who inspects each cargo.

A snag in raising the minimum weight is that it was reduced only in July from 200lb to 110lb. At 200lb an average Friesian calf is ten to 12 weeks old, which is around the age for slaughter for white veal. The weight was reduced in response to pressure from exporters only after trial cargoes had been accompanied by Ministry vets. They found that 110lb calves, aged from seven to 14 days, suffered no ill effects on short overseas journeys.

The 110lb rule applies only to journeys up to 200 miles. This limits it to Belgium, Holland and France. During the recent export boom, however, calves shipped or flown from Kent to Ostend have been ferried on to a number of countries, including Italy.

Calf-traders find a new place to sell
John Winter 18.10.68

Confusion grew yesterday in the campaign by animal welfare workers and beef farmers to curb the "white veal" trade in young calves between Britain and the Continent. The Belgian Government's decision – reported exclusively in the *Daily Mail* yesterday – to refuse entry to calves weighing less than 176lb was apparently not known to one export centre – Lydd Airport, Kent. A consignment of 180 calves left in a Bristol freighter for Ostend, Belgium.

All were above the British minimum weight – 110lb – but an official of the airline, British Air Ferries, said he had not heard of the Belgian restriction. The Ministry of Agriculture also said yesterday that it did not know "officially" of the new Belgian rule. Even with the Belgian clamp-down on the calves which are reared on an unbalanced diet to make their flesh white, a new export market opened up yesterday – in France.

A cargo of 150 calves was shipped from Weymouth, Dorset, to Le Havre. This is said to be the first instalment of orders totalling 40,000. The buyers – a group of French farmers – have promised that the calves are for beef rearing and not for the "white veal" slaughter yards.

Mr Alick Buchanan-Smith, Tory MP for North Angus and Mearns, yesterday put down two questions for Mr Cledwyn Hughes, Minister of Agriculture, to answer in the Commons on Monday. One asks how many calves have gone to Europe in the past three months and what effect the exports have had on home supplies of stock. The other concerns imports of meat from overseas produced on rearing systems which are not in line with the Ministry's recent draft codes of practice for factory farming.

Mr Hughes, who told MPs on Wednesday that the whole "calf drain" is under urgent consideration, is now expected to put a brake on the trade, which has exceeded 3,000 animals a week.

Ministry experts are believed to be considering a limit of 175lb – almost the same as the new Belgian one. This would mean that export calves, which at the present limit of 110lb are two or three weeks old, would be eight to ten weeks, which is approaching the age for slaughter on the white veal system. They would also have been fed for several weeks on solid food which would darken their flesh.

The great 'calf drain' muddle

Rhona Churchill and John Winter 2.11.68

The *Daily Mail* now has irrefutable evidence that roughly 29,000 young bull calves from British farms were exported to Belgium during the last three months. This figure was given by Dr Jean Pierre Latteur, chief veterinary inspector for the Belgian Ministry of Agriculture, who supervises the import into Belgium of all young calves.

It is 9,000 in excess of our original estimate, which was hotly denied by the Ministry of Agriculture. It is 21,000 in excess of the Ministry's estimate. And, it is in direct conflict with the Ministry's statement in the Commons on October 21 that such shipments were "very little higher" than at the same time last year, when they were a mere 7,500.

The discrepancy between estimate and fact is so vast as to be almost inconceivable. Yet, our new figure comes from shipment records compiled at the Belgian ports of entry and passed on to the Ministry for their official records.

Faced with these new figures yesterday, a Ministry of Agriculture official said: "Our returns indicate that the number of calves exported up to mid-October was about 8,000. If there is other reliable information which differs from this, the Ministry would be glad to study it."

Has there then been large-scale smuggling of calves out of Britain? Or are the returns of our Ministry's vets seriously incomplete?

Our beef farmers have known only too well what was going on. They have been going to the markets in search of cattle for fattening for beef, but have been outbid by exporters willing to pay £5 to £8 more than the normal rate per calf. Our shippers have known what was going on. They have been moving calves to Ostend at the rate of up to 900 a week.

The RSPCA has known. It has been alleging cruelty and the fact that most of the calves will be reared for white veal on the type of unbalanced diet now condemned in Britain.

The deeper one delves into this great calf drain muddle, the more curious it grows.

Calf 'king' on the move

Harry Longmuir, Ostend 18.11.68

Mr Douglas Clay, British cattle "king" who has shipped about 8,000 young calves to Belgium since mid-September, is branching out into France. He has been responsible for more than half the total British exports of calves for veal to Belgium.

Mr Clay, of New Barn Farm, Hooe, Sussex, said there today: "The trade to Belgium is drying up. There looks like being a switch to France. A lot have recently gone there from Weymouth, and I have been talking business with the French.

"What's all this fuss about some calves being exported? I have done nothing wrong. It's all approved by the Ministry of Agriculture. All right, a minority of

British farmers are complaining, but most of them are quite pleased. There are more farmers selling calves for export than there are farmers buying them for beef production. Anyway, I wouldn't say the prices for calves are much more than £2 or £3 higher than they would be if there was no export trade. I'm not making a pile of money out of this, you know. In fact, recently I've been selling at a loss."

So why continue?

"I am involved in buying stock, owning lairages (box enclosures for cattle), labour, boats, exporting. You can't always make money on everything. Recently, by the time I have paid duty and costs I haven't been making anything on exports to Belgium."

Mr Clay ships calves bought by his agents all over England from lairages at Sheerness, which he owns, in boats which he also owns. He also owns three farms in Sussex.

"I don't go to markets very often. I concentrate on the selling. I'm over here every ten days or so, and leave the buying to my agents. I think I have shipped about 6,500 calves up to the middle of October, and I've done nearly about another 1,500 since then. I can't say why there should be a disparity between the Belgian and British figures. You can forget about smuggling. It is impossible, and not worthwhile, anyway."

Had large-scale buying for export resulted in a shortage of calves for fattening, and would it cause a serious shortage of home-raised beef in ten to 15 months?

Mr Clay replied: "I can't speak for the beef producers. But I have had no difficulty in getting livestock. I have never known any time since rationing ended when there was any shortage of beef in

the shops at prices lower than in any other country."

Did he think the opponents of calf exporting had a case either on economic or humanitarian grounds?

'If the British Ministry of Agriculture says it's all right, as it has done, who am I to argue? If it could see anything wrong about it the Ministry would ban it.'

John Winter writes: Mr Clay's up-to-date figure of 8,000 calves, plus those sent by other exporters, carries the total exports to Belgium far above the "official" Belgian figure of 10,562 and the Ministry's own figure of 10,757 given in the Commons on November 7.

The whole question of calves for export is being raised in a Commons debate tonight by Mr Peter Mills, Tory MP for Torrington, Devon. Another Tory, Mr Alick Buchanan-Smith (North Angus and Mearns) has tabled two questions demanding facts and figures.

Exit the Little Lion – enter higher prices   *John Winter 20.12.68*

Who mourns the Little Lion, controversial symbol of the Egg Marketing Board? After exciting more passion and fury than any other trademark since he first appeared on eggshells 11 years ago, he was killed officially and predictably by Mr Cledwyn Hughes, Minister of Agriculture, yesterday. I believe his passing will be regretted by more people than the Minister imagines.

To start with, millions of housewives will be faced with a choice of eggs which all look the same. (The only stamped eggs likely to be around after Monday will be very limited and erratic supplies from abroad, which must still be stamped "foreign" or with their country of origin.)

I forecast that more than half the housewives of Britain who have bought lion eggs will, if they continue buying the same kind, have to pay more for them. Why? Because with the stamp removed, demand for the board's eggs will exceed supply.

Housewives will also face a bigger risk of getting stale eggs or eggs of uncertain quality from unscrupulous suppliers.

More fortunate will be those who always buy so-called "farm-fresh" eggs – without the stamp. These have usually cost from 6*d* to 1*s* a dozen more than lions. Now "farm-fresh" eggs will find it hard to maintain a premium over ex-lions on the same counter because to the shopper they will look exactly alike.

 # EMPLOYMENT

Chance to learn new skill

Peter Bullen 17.1.68

Crash refresher courses covering several forms of farming are being arranged for victims of the foot-and-mouth epidemic, the Agricultural Training Board said yesterday. Details will be published by the end of January about most of the three-day courses to be held in the infected counties.

Mr Richard Swan, the board's chief training adviser, said some farmers would probably take the opportunity to change completely to new farming enterprises. The courses would enable farmers and workers to refresh their knowledge and learn any new skills and techniques which might be necessary to cope with new situations.

'Sue me' farmer defies training board

John Winter 22.5.68

Colonel John Baker White, 66, a Kent farmer, yesterday challenged the Agricultural Training Board to sue him in the county court for a £12 board levy. He said: "I have written to the board that I am ready to be taken to the county court so that the decision of the court can be obtained."

He says that the board has no legal power to recover the money from him because the words "recoverable, creditor, debtor" are not included in the statutory instrument which authorised the levy.

Mr Baker White, who employs four farmworkers, said: "My wife and I do not intend to pay the levy until the whole character of the board and its budget are checked, or until we are ordered to do so by the court."

Scrap training board call by farmers

Peter Bullen 20.9.68

The National Farmers' Union suddenly rebelled yesterday against the Agricultural Training Board that has been in existence for more than two years. The union's council asked the Government to scrap the board completely and set up another scheme, tailor-made to suit the industry and paid for by the Exchequer, at the annual farm price review.

Only five of the 140 council members were against the proposals, which were sent to Mrs Barbara Castle, Minister for Employment and Productivity, and other Government Departments. But Mrs Castle, who warned the industry only two weeks ago that it must support the board, sent back a prompt, frosty reply that legislation could not be changed just to suit farmers. But she said she would consider the NFU's suggestions.

The NFU's decision to call for the scrapping of the board, which the union itself asked the Government to set up, comes after two years of bitter wrangling.

Opposition has been so great that even now the board has not collected all its first-year levy money from farmers. More than 50 per cent of the 98,000 employers have not yet paid the £3-a-worker levy and more than £500,000 is outstanding.

After weeks of discussions with the board, the NFU managed to get it to trim its original plans that would have cost farmers a levy of £6 a worker, but half its members would not accept this compromise.

Mr Bill Williams, NFU president, said yesterday that further discussions the union had been having with the board and Mrs Castle's Ministry had come to a dead end.

The Industrial Training Act, under which the board was set up, was just not suited for agricultural training problems. He said that the choice was either a fresh start with a new scheme or massive non-cooperation from a large section of the industry.

Mr Williams stressed that farmers still wanted improvements in training but in a scheme that came under the agricultural departments and financed from the price review, preferably in the form of a production grant. Local training facilities would be used to the maximum.

When the union originally requested a board it had not realised that its administrative costs would be so high.

Rebel farmers say: "We'd rather go to court"
Peter Bullen 3.10.68

Hostile National Farmers' Union members have started to rebel against their leaders' advice on the Agricultural Training Board. The NFU told the Government that the Board was not wanted by farmers and should be scrapped. But it advised its members to pay the £3-a-worker levy due to the Board for its work last year.

So far less than half the 98,000 employers in the country have paid the levy and more than £500,000 is outstanding. Yesterday Surrey NFU Horticultural committee said: "We remain firmly of the opinion that it would be wrong to pay any levies to the present Board. Members should not pay any levy until served with a summons."

County secretary Mr Philip Shaw said all the growers on the committee had backed the move and were urging members throughout the county to follow their lead. One committee member, Mr Robert Hewitt, said that if he succeeded in withholding his levy payments of £234, he would distribute them among his workers as a bonus.

Last night NFU headquarters said its only comment was that there was a duty on farmers to pay the levy to comply with the law. A training board spokesman said withholding levy payments until a summons was issued only caused extra administrative work and extra cost which meant less money would be available for training.

North Lincolnshire farmers are organising a public meeting opposing the board. About 40, who have received solicitors' letters asking them to pay their levies, have decided they would rather go to court.

Tight-fisted
John Winter 28.10.68

A woman trade unionist accused farmers of being "the most tight-fisted employers in any major industry today" during a weekend conference at Hereford.

Miss Joan Maynard, vice-president of the National Union of Agricultural Workers, said farmers were bang up to date when it came to new techniques to improve the industry. "But their approach to wages is eighteenth century," she claimed.

No pay-out until February – No cut in 44-hour week – Farm workers to get 17s rise
Peter Bullen 20.11.68

Farmworkers were awarded a rise of 17s yesterday, bringing the minimum weekly rate for men to £12 8s. It is far below the £4 9s increase they claimed, and there is no cut in the 44-hour week. The rise will be paid from February.

Lord Collison, general secretary of the National Union of Agricultural Workers, said: "It is a disappointing award. I think it will do a little to stop the drift from the land, but it is not enough to stop it."

The award would do little to close the gap of nearly £7 between farmworkers' earnings and those of workers in other industries. Lord Collison said it would be difficult for farming to expand and save food imports unless it retained its workers.

The 17s awarded by the Agricultural Wages Board for England and Wales represents an increase of 7.3 per cent and is the biggest ever for farmworkers. There will be proportionate increases for women and young workers, and the overtime rate will go up 7d to 8s 6d an hour.

The union had asked for a four-hour cut in the 44-hour week. Throughout yesterday leaders of Britain's 383,700 farmworkers argued with farmers' representatives on the Board about the hours issue. But after eight hours' deadlock, one of the longest meetings on record, the farmers' proposal of 17s extra and no cut in hours was carried by the board on the votes of its independent members.

Lord Collison said: "It has been a long and tedious day."

The union executive will meet next week to decide what action to take over the award.

Mr Henry Sharpley, leader of the National Farmers' Union representatives on the board, described the award as a "very, very considerable increase". He said it would cost the farmers an extra £17 million in a full year and, with the overtime increase, give farm workers £1 a week more. Mr Sharpley said that horticultural producers and farmers whose commodities were not covered by the annual farm Price Review would have particular difficulty in recouping the extra cost as they could only do this through higher market prices.

They had suggested the 17s offer against the background of the Government's statement last week about encouraging the import saving. This would require workers at least to maintain their present hours, so they had opposed any attempt to cut the working week.

The previous biggest award made by the board was the 15s given last February.

Letter of the week – pay award deserved
(Mrs) M M Brown, Abbots Bromley, Rugeley, Staffordshire 26.11.68

Three cheers for the farmworkers and their peaceful march to the singing of

The Farmer's Boy. They deserve their pay award of 17*s* a week. It was my privilege to work with them on my father's farm for many years. Surely this fine body of men should have their increase backdated to July, like the Post Office workers, instead of having to wait until February. They deserve a Christmas box, too.

Barbara may freeze farm-workers' rise   *Peter Bullen 24.12.68*

The 17*s* pay rise awarded to 380,000 farmworkers may be frozen by the Government.

Leaders of the agricultural workers' union refused yesterday to accept a Government plan to pay 10*s* from February 3 and refer the rest to the Prices and Incomes Board. And now Mrs Barbara Castle, Minister of Employment and Productivity, may stamp on the whole 17*s*.

Lord Collison, union general secretary, said after meeting Mr Roy Hattersley, Mrs Castle's Joint Parliamentary Under-Secretary: "I am appalled and disgusted." He said the 7⅓ per cent award was fully justified under incomes policy because farmworkers were among the lowest-paid in the country. And their productivity record has been running at more than 6 per cent a year since 1961.

Mr Hattersley said the Government endorsed an increase up to the 3½ per cent maximum, but wanted to discover if the rest of the award was justified on productivity grounds. The rise, proposed by the Agricultural Wages Board, would bring the minimum wage for farmworkers to £12 8*s* for 44 hours.

Barbara Castle relents on farm pay freeze   *John Winter 31.12.68*

Mrs Barbara Castle yesterday withdrew her threat to freeze the 17*s*-a-week pay rise proposed for Britain's 380,000 farmworkers. Instead the Minister of Employment and Productivity came up with a compromise solution.

Her earlier plan, that the farmworkers should take 10*s* now and have the other 7*s* investigated by the Prices and Incomes Board on productivity grounds, met with determined opposition. She has now referred the whole question of farmworkers' wages to the PIB for a crash investigation before February 3, when the 17*s* a week rise should come into force. It means that once again Mr Aubrey Jones, PIB chairman, is being forced to arbitrate on the Government's incomes policy.

But for the farmworkers, who were threatened with a three-month freeze on the whole 17*s* increase on their £11 11*s* minimum wage for a 44-hour week, the compromise solution is an 11th-hour victory. Tomorrow the statutory Agricultural Wages Board meets to ratify its proposal to raise the minimum by 17*s*.

If the PIB says the full 7⅓ per cent increase is justified it will be paid from February 3. If not, the Government will look at the whole situation again. There is little doubt that the PIB will be able to finish the examination by February 3, as it did a lot of the spadework in an investigation of farmworkers' wages two years ago. The farmworkers argue that they work longer hours than most other industries, but earn on average £6 a week less, although their productivity record is one of the highest in the country.

Yesterday the National Farmers' Union told the Government that the pay award was consistent with the Government's incomes policy and should not be referred to the PIB.

PEOPLE in the NEWS

Peart told to 'face music' in milk row
John Winter 15.2.68

Mr Fred Peart, Minister of Agriculture, was criticised by the Law Lords yesterday and told he "must be prepared to face the music in Parliament".

They were giving judgment against him in a case about the Minister's refusal to investigate a complaint over milk prices. Three dairy farmers, on behalf of 4,000 others in the South-East, have fought 11 years to establish that they are forced by the Milk Marketing Board to subsidise farmers in other regions. They took the Minister to court because he refused to set up a committee of investigation into their complaint. The High Court found for the farmers, and the Appeal Court, by a majority, for the Minister.

Yesterday, the House of Lords Appeal Committee, by a majority of four to one, allowed the farmers' appeal and directed Mr Peart to consider their complaint. Lord Upjohn said he had not given a single valid reason for refusing to order an inquiry. A Ministry letter showed a scarcely veiled fear of parliamentary trouble and its possible results if an inquiry was ordered.

Lord Pearce said that the South-East region was in a minority on the board and had been unable, despite 15 attempts, to persuade the majority to do anything about it. In the populous South-East milk was more valuable. The consumer was nearer and cost of transport less, but overheads were generally higher. Land was more expensive. It seemed to follow that if a milk producer in a populous region was paid the same as a producer in a sparsely populated region he was not being fairly treated.

The Lords directed that farmers should have their Court of Appeal costs and two-thirds of their House of Lords costs.

The farmers, Mr George Padfield, of Chambers Farm, Epping, Upland, Essex, Mr Geoffrey Brock, of Blackwell Farm, Compton, Surrey, and Mr Henry Steven, of Westerham, Kent, all listened to the judgment and then joined in a celebration lunch with Mr Eric Quested, regional board member throughout the fight.

Twice during the expensive legal battle farmers throughout the region backed a fighting fund at the rate of $1s$ a cow.

Peter Bullen commended
John Winter 15.1.68

Peter Bullen, assistant agricultural correspondent of the *Daily Mail*, has been specially commended in this year's Fisons Awards for outstanding agricultural journalism. The judges said that his news feature on Stansted, in the *Daily Mail* of June 15, 1967, was a powerful presentation of the farmers' case in a major dispute.

The main prize, a travel scholarship of £300, was won by Michael Williams, of the *Farmers Weekly*.

Sir Richard boycotts BEA for using instant milk
Peter Bullen 14.5.68

Sir Richard Trehane, Milk Marketing Board chairman, has asked his colleagues and staff to boycott British European

Airways because it uses skim milk powder on many of its flights. Sir Richard said yesterday: "My fellow board members, the staff and I will use other airlines wherever possible. It is the only action left to us."

Board members and officials make dozens of flights a year. Sir Richard estimates that he has travelled well over half a million miles by air since joining the board 21 years ago.

The fight to get BEA to use liquid milk instead of instant milk powder began about 15 months ago. Sir Richard said he wrote to the BEA chairman, Sir Anthony Milward, complaining about the powdered milk. Although he had been told a change to liquid milk and fresh cream would be made, BEA was still using instant milk powder.

Sir Richard said: "I have written letters and tried three times to get Sir Anthony Milward to meet me to discuss this over a drink, without success."

The board would be glad to supply milk and cream in any containers to suit BEA, he said. A board official said they were upset about the use of instant milk powder as it was not real milk, only a substitute. He added: "We supply fresh milk to Quantas, Air France, TWA, Air Canada, PanAm, El Al and Lufthansa for outward flights from London, and they all use cream."

BEA said: "We serve fresh milk on all domestic flights: fresh milk and cream on all first-class flights and powdered milk only on international tourist flights. The reason for this is that we carry so many passengers – about 7 million a year – that we must give a good service."

If stewards and stewardesses spent a lot of time doling out small, individual quantities of milk it could cause delays and difficulties. In some cases it was also difficult to pick up the high quality milk that BEA wanted. A litre bottle of fresh milk was always carried for children and babies.

Who said British wives can't cook! *John Winter 28.5.68*

Mrs Kathleen Thomas has just trebled the size of the kitchen at her Devon farmhouse by converting and taking in the cowshed. With three cooks beavering away in it and 15 housewives cooking like mad in their own kitchens, she is just about keeping pace with the eating revolution she started simply by offering only home-made food.

She is selling through her own two town-centre shops. She has the catering at a Taunton hotel, runs the food service at Exeter's new repertory theatre on the university campus, and has just accepted the lease of a major licensed restaurant in Exmouth. Seventeen other hotels have offered her their catering contracts. A brewery chain and a national stores group are clamouring to have her food on their counters.

Some going for an attractive grandmother. A farmer's wife for 22 years then a cookery writer for ten, she went back to Devon to explode the myth that the affluent, sophisticated English are irretrievably sold on Continental food.

After weekly extolling in a farming journal the virtues of farm food cooked in farm kitchens, Mrs Thomas came to regard her new venture as a missionary enterprise. It has quickly grown into a sizeable and rapidly expanding business, with her husband Roy, retired from farming, as business manager.

Mrs Thomas's revolution started in a roadside café on the A30 near Honiton four years ago with an annual turnover of £9,000. Last year, for only half of which she had the Taunton hotel, it was £47,000, and the staff had grown to 20-odd. Her butcher's bill, all prime West Country beef, lamb and pork, now tops £100 a week.

The formula is simple. She buys only high-grade local produce. She concentrates on traditional West Country dishes, interspersed with ingenious recipes she has invented. They roll off the tongue and melt on it – harvest pie, country terrine, Somerset slice, cheese eggs.

Kangaroo pigs   *John Winter 24.7.68*

Breeder of champion pigs, Mr Charlie Flack, of Suffolk, said yesterday that the Pig Industry Development Authority's boar-testing scheme would lead to pigs which looked like kangaroos. He told the annual meeting of the National Pig Breeders' Association in London: "We are paying money to PIDA to have these boars tested and what a parcel they are. Some of them are shockers."

Other breeders were equally critical of the type of boar which was earning high points in the testing station.

GENERAL

In goes muck, out comes money   *Peter Bullen 3.1.68*

Tons of rubbish – old carpets, curtains, bicycles, furniture, even kitchen sinks – may soon be turned into cash by a new machine. The Government-financed National Research Development Corporation plans to back it with £150,000.

The new process will turn the rubbish into profitable bags of fertilisers for gardeners and farmers. No sorting is needed because materials not needed are automatically rejected.

Britain produces ten million tons of rubbish a year. Local authorities spend £40 to £50 million a year on refuse disposal. The new process gives them a chance to make a profit. The plant, developed by the Lawden Manufacturing Company of Wolverhampton, will produce "Sweetsoil", a soil conditioner, to bind sandy soil or break up heavy clays, or with added chemicals, be sold as a fertiliser.

The cost for a town of 100,000 is estimated to be about £350,000 to build the plant on a third of an acre site. This would produce about 300 tons of fertiliser a week, which could bring in about £150,000 a year. Dr Basil Bard, a chief executive of the development corporation, said they were prepared to provide about 50 per cent of the cost of installing the first plant. If it was a success they would get their money back.

Proving that you can't keep pigeons on the Pill!

John Winter 27.3.68

Scientists have been feeding pigeons with a contraceptive pill in an effort to reduce their numbers.

You may think this is a cynical trick to play on the bird which originated the delights of billing and cooing, but complaints about the damage and nuisance which pigeons cause are worldwide and growing. In this country it is the ravages of wood pigeons on farm crops. In the US, which has no wood pigeons, their town cousins are showering more and more debris on cities, and it is there that scientists have been experimenting with the pill.

The results are described as "very inconclusive", which in lay terms means that it does not work. This may be because the scientists do not know enough about the love life of pigeons, or that they have found a contraceptive chemical which the pigeons can be kidded into swallowing. The trouble is that you can never be sure the birds have taken the pill in time.

Anyhow, our scientists will have none of it, but as they are being badgered by angry farmers who are losing more money each year, they have got to try something. In the past few months pigeons in unprecedented numbers have ravaged winter crops, especially Brussels sprouts. The main factor is probably a rather mild winter, the third in succession, in the important sprout growing areas, which allowed many extra pigeons to survive. Farmers also blame the ban on shooting during the foot-and-mouth epidemic.

A survey by the National Farmers' Union in Bedfordshire and Huntingdonshire, two counties with big acreages of vegetables and cereals, shows the cost of pigeon damage and attempts to control it over the past 12 months as £750,000. It must run to many millions throughout the country. Two-thirds of the loss in the two counties was on sprouts – £30-an-acre damage and £9 on control measures.

Another £73,000 went on cabbages and £49,000 on peas, to which pigeons are partial at the seed and seedling stages, after the winter crops have been cleared. The birds became so tame that they almost ignored such old-fashioned scarers and the kind that go "pop" at regular intervals.

Today an NFU deputation will descend on the Ministry of Agriculture to demand action. They will claim that the only solution in sight is another device with which Ministry boffins have been experimenting for five years – narcotics. The idea is to coat a pigeon's favourite tit-bit with dope. Then, while the birds are reeling round, they are humanely killed.

The process raises the hackles of bird protectionists. What about other, smaller birds which might eat the doped bait with fatal results? Or the pigeons which are not picked up? Game-bird enthusiasts are also sceptical. What will go down a pigeon's gullet will also fit into a pheasant's, and what could be more tempting to a poacher than a cock-eyed pheasant ready to stagger into his hands?

The Ministry men know this, which is why they are so reluctant.

They have perfected the right dope, alpha chloralose. It will stupefy a pigeon in 15 minutes, less if it is greedy, and keep it "out" for a time which varies with the amount consumed. They have also selected the bait – tick beans. They are too big for a small bird to swallow (but not a pheasant or partridge).

They have proved in their own tests that the idea will work – provided there is an adequate and experienced mopping-up squad to collect the stupefied birds. They doubt, however, whether the average farmer could muster such a staff. Cautiously this year the Ministry will offer special licences to a few farmers to use narcotic bait on a trial basis. Ministry experts will supervise and report on the results.

I regret to warn the suffering farmers

that I have never known the Ministry to approach any kind of trial with so little enthusiasm.

It's a knock-out . . . farmers wage dope war on pigeons

John Winter 27.5.68

Two farming neighbours have collected 170 unconscious wood pigeons, two crows, one turtle dove, and one skylark from two fields of brussels sprouts in a fortnight. All had been knocked out by eating doped tick beans supplied by the Ministry of Agriculture for trials on the use of stupefying bait to reduce the havoc caused by pigeons on brassicas and other vegetable crops.

The pigeons were painlessly destroyed while still unconscious. So were the crows. The dove and lark both recovered after being kept warm until the following day.

Both farmers, Mr Maurice Clarke, of Silsoe, Bedfordshire, and Mr Dick Woodward, of Maulden, are delighted with the tests and will ask the county NFU branch to press the Ministry to make the knock-out drops available under licence to all farmers who suffer pigeon damage.

Twenty-six farmers are testing the method in Bedfordshire, and Ministry observers are keeping records of the daily toll of pigeons on each of 42 sites. This is the result of five years' investigation by Ministry scientists, who have been extremely cautious in permitting the use of dope because of the risk to other birds, both smaller wild species and game birds, chiefly pheasants and partridges.

The dope, alpha chloralose, will knock out a pigeon for up to eight hours, according to the amount consumed, after which it will recover if not exposed to severe cold. It is applied in an oil coating to the beans, which are too big for a small bird's gullet but not a game bird. If the weather is dry the treated beans will remain potent for two weeks.

Last year Mr Clarke lost £1,500 through pigeon damage on 100 acres of brassicas, and Mr Woodward suffered £900 damage on 33 acres. Mr Woodward took me across a two-acre field of young sprout plants, many of which had been shredded or eaten down to the stem by pigeons, before they turned to the beans scattered on the bare earth. He said: "These plants are beginning to recover now we have cleared so many pigeons."

Letter of the week – stock doves

N R Went, Eccleshall, Staffordshire 3.6.68

I read with interest that doped tick beans are being used against wood pigeons. This raises many problems, not least that of accurate identification. How many farmers would be able to identify with certainty a stock dove? This species has never been numerous and we can ill-afford to lose the few that remain. I hope such points will be considered before the dope is made available to farmers.

1969

The political front was the battle ground for adequate compensation for foot-and-mouth disease victims and the £ for £ quarrel lasted for several months. Farmers again feared the worst when the Price Review negotiations took place with growing threats of disruption and industrial action. The control of farm chemicals was under discussion and the wet winter threatened the corn crops. Anti Agricultural Training Board fever was at its height fuelled by the outstanding problems of levy payments. There was a new deal for "tied cottagers" and the workers got a record pay rise. Lord (Harold) Collison, former workers' union leader called it a day and John Winter won the Fisons award for outstanding journalism.

 ## POLITICS and FOOD

Whitehall 'threat to family farmers'
John Winter 7.1.69

If Government policy is not changed the countryside will be transformed within ten years into a kingdom of huge factory-style farms in which surviving family farmers will face a growing struggle for existence, it was claimed yesterday.

An appeal to Whitehall to recognise that it is forcing medium-size owner-occupiers to give up farming and farm-workers to seek work elsewhere was made by Mr Charles de Boinville, chairman of the Unilever UK Animal Feeds Group, at the Oxford Farming Conference dinner last night. He said that by 1978 dairy farming would be so concentrated that one-third of the nation's cows would be in herds of over 100 with at least 50 herds of over 1,000. The number of dairy farms would fall from 95,000 today to 55,000.

Three-quarters of all laying hens would be in flocks of over 20,000, with probably five of over one million.

Future policy should aim at maintaining a varied, profitable and broadly based agriculture.

Letter of the week – working all hours
(Name supplied) Salisbury, Wiltshire 3.3.69

I am sure my husband and I are in the same position as many tenant farmers: working all hours of the day and quite a few at night tending stock. After the disastrous harvest we both have had to take additional jobs. This means the children have to get themselves off to school and often come home to an empty house.

We have fought before, but now we are fighting desperately hard. Is it worth it? The farm price review gives little encouragement. Is the Government out to break the smaller man in any business with faith in his own ability?

MPs blame Board of Trade over farm imports   *John Winter 6.3.69*

The Board of Trade is the chief stumbling block to expansion of home food production to replace imports, an all-party committee of MPs said yesterday. The 25-member Select Committee on Agriculture also criticised the Department of Economic Affairs and the Treasury for playing down agriculture's potential.

The Committee, set up as an experiment two years ago, was wound up by the Government last week. It is to protest vigorously over its fate.

In its report on the possibilities of expanding agriculture to save imports it says that farmers should be encouraged to step up production not just to keep pace with rising food demands, but to replace some imported supplies. The committee says this is technically possible if farmers receive adequate cash incentives from the Government, the drift of agricultural workers is halved, and expansion is not tied to rigid international agreements.

But, says the report, Board of Trade witnesses made it clear that in any conflict exports took precedence over import saving.

Wreath for a piece of democracy   *Peter Bullen 13.3.69*

Twelve MPs, most wearing black ties, met in a committee room at the Commons yesterday for the "funeral" of a piece of parliamentary democracy. On the table was a wreath with the message: "In loving memory of the Select Committee on Agriculture, which died in its prime."

This unofficial ceremony was the way members of the committee chose to launch their final report to the House – a report in which the 25 MPs from all parties pilloried the Government for the way it killed off the committee only 27 months after setting it up.

And in six short recommendations it outlined how it thought specialist select committees could contribute to the work of Parliament. The measures were: that the committees should have a measure of permanence; they should be consulted over changes in their size, membership and duration; they should be limited to about 16 members and be adequately staffed; and they should not be inhibited by a Government from considering policy in its formative stages.

Farmers ready to revolt as they hear worst   *John Winter 20.3.69*

The threat of a farmers' revolt came yesterday after details of the Government's "cheese-paring" annual price review were revealed. Farmers were recommended by their leaders not to co-operate any more in the home-produced food expansion programme without considering if it is worthwhile. This warning of revolt was made by 200 National Farmers' Union council members and county secretaries.

The settlement gives farmers only £34 million increased support in the face of a £40 million rise in costs and £39 million drop in income through last year's wretched harvest. Mr Bill Williams, NFU president, described the review as "cheese-paring and penny wise and pound foolish." It had missed the opportunity of slashing the £1,000 million-a-year import bill for food we could produce at home.

But the NFU went further in a resolution. First it rejected the settlement for failing to give farmers the necessary money to expand or the assurance that imports would not be allowed to undermine markets. Then the council accused the Government of letting down thousands of small family farmers who will be critically hit.

After warning all farmers to reconsider their future production, the resolution ended: "The result is bound to undermine any remaining confidence farmers had in the reality and sincerity of the Government's agricultural policy."

The union's disillusionment is intensified by figures in the review White Paper which show that because of the 1968 harvest, farmers' output slumped by 3 per cent while their brilliant record of rising productivity, 6–7 per cent a year, was halted.

The £34 million award, as I forecast in detail on Monday, is concentrated wholly on products earmarked by the Government for expansion to save £160 million a year in imports in four years. These increases, says the Ministry of Agriculture, represent an injection of £37 million in the selected group of import savers.

But the flashpoint for revolt and bitter accusations of betrayal by 90,000 dairy farmers is the "nil" award for milk.

End of 3-year honeymoon

John Winter 20.3.69

This year's price review marks a watershed in Government agricultural policy by ending the three-year honeymoon during which it has accepted that it could not get more home-produced beef without a build-up of unwanted milk. Yesterday the Government asked plainly for

more beef without more milk. And the milk price would have been cut if assurances had not been given in the past three reviews that higher milk production would be accepted.

The White Paper says ominously: "Many farmers will need to change their husbandry pattern and practices."

This is a bitter blow for thousands of small dairy farmers, mostly in the western counties, who depend almost entirely on milk and have little chance of switching to other products. They face an inevitable increase in production this year, probably 80 million gallons, because of the rise in the number of cows. This can only lead to a further drop in their "pool" price, which takes account of milk sold cheaply for butter, cheese and other products. This price is already down by 1.4d a gallon this year.

Farmers are also furious that the White Paper still talks about the theoretical £30 million value of their increased productivity in a year when they are £39 million out of pocket and productivity is expected to remain stationary.

The White Paper claims confidently that its concentration of extra cash on import-saving products will provide a "solid base for advance by the industry as a whole and, with other measures, should set the industry on course. Based on the underlying strength and resilience of an efficient industry they should ensure recovery, expansion, productivity and profit."

The new guaranteed prices are:

Beef cattle, 215s a live cwt (15s up); *sheep*, 3s 7.75d a lb (1.5d up); *pigs*, 48s 5d per 20lb deadweight (6d up); *milk*, 3s 9.26d a gallon (0.4d up).

Potatoes 302s 6d a ton (5s up); *wheat*

29s a cwt (1s 7d up); *barley* 26s a cwt (10d up); *oats* 27s 10d a cwt (same); *rye* 21s 7d a cwt (same); *sugar beet* 136s 6d a ton (same); *wool* 4s 5.25d a lb (same); *hen eggs* 3s 6.07d a doz (1.19d down); *duck eggs* 2s 5.07d a doz (0.38d down).

The prices for barley and pigs are further improved by the abolition of the limit on the amount of barley qualifying for guarantee and a big extension in the number of pigs qualifying.

● Mr Cledwyn Hughes, Minister of Agriculture, told a Press conference yesterday that he believed the price review was a good one for the housewife.

Milk prices would not go up and increased supplies of home-produced beef and pork would help to keep prices steady.

Quit now, Tory tells Hughes

Peter Bullen 22.3.69

Tory agriculture spokesman Mr Joseph Godber yesterday called on Minister of Agriculture Mr Cledwyn Hughes to resign. British farming had taken a terrible blow to its confidence, Mr Godber said at Scarborough.

"One part of the tragedy is that the Government as a whole and the Prime Minister and Mr Cledwyn Hughes in particular should have brought political integrity so low. In the past four years British agriculture has been bled white. Capital and confidence have been slowly ebbing away. A major effort will be required to stop haemorrhage and nurse the industry back to health. Only then can we turn to consider import savings again."

The price review pin-pointed a number of lessons. It showed the Government's total lack of interest in British agriculture, whose loyalty and enthusiasm had been cynically exploited. It had also established beyond question the thesis that the Treasury would never make expansion possible by providing sufficient cash for adequate guaranteed prices.

Letter of the week – a source of embarrassment

Constance Pirie, Woodbridge, Suffolk 24.3.69

When will farmers realise that neither this Government nor any other Government want the food we produce?

The farming community is a source of embarrassment to any party which happens to be in power, and the public are totally indifferent to farming.

The only worthwhile crop to be grown on land now is houses.

Britain gets Mao butter in disguise

John Winter 31.3.69

British housewives are being sold Chinese butter disguised under "blended" labels. It is being imported at a time when our warehouses have record stocks of 110,000 tons of butter. The butter stocks are so large, in fact, that on March 20 the Board of Trade slashed by 65,000 tons the import quotas for the next 12 months of 17 out of 19 overseas suppliers.

But China does not figure on the official list of countries whose shipments are limited by tonnage quota. Chairman Mao's butter comes under a long-standing trade agreement between Whitehall and Peking which gives him an allocation worth £100,000 a year. It was not even mentioned in the Board of Trade quota announcement.

More than 250 tons has been shipped in the last six months, and its arrival has infuriated British dairy farmers whose milk price has been undermined by excessive imports of varied products. The first complaint about these "phantom" imports came from Mr Sid Pennington, a dairy farmer from Cobham, Surrey.

In a local hardware store he spotted a box labelled on one side, "Butter from the Chinese People's Republic". Mr Pennington was given the empty box and he passed it to the Surrey county branch of the National Farmers' Union.

I traced the importers, a London firm. An official said: "We handle most of the butter from China. It comes in 56lb bulk cartons and is of reasonable quality. It is used for blending; it is a nice, quiet little trade."

The Board of Trade said that China's £100,000 allocation was granted under an agreement made before the quota system was established in 1962.

The Milk Marketing Board has established that from October to February three shipments of Chinese butter totalling 252 tons reached Britain. The landed price averaged just under £262 a ton which means the £100,000 will cover about 380 tons. This is a minute fraction of the total imports this year – at 460,000 tons – but two countries on the quota list have smaller tonnages.

Letter of the week – dumping

D Lacey, London W9 7.4.69

I was interested in your Farm Mail report on butter imports from China. I enclose a wrapper from a tin of Chinese green beans that were of quite fair quality, but the price of 1s 11d for a can purporting to be 1¼lb indicates dumping. Personally, I would not have bought them: someone else did the shopping. Surely, in view of the outrageous treatment of at least one

of our nationals by Peking this trade should be smartly cut off?

A bob on everyone's food bill

Peter Bullen 7.5.69

Household food bills went up by 1s a person a week last year – to 37s 11d compared with 36s 11d in 1967, a Government survey showed yesterday. Food prices averaged nearly 4 per cent higher than a year previously, although there were small decreases for a number of commodities, including cheese, butter, potatoes and some kinds of fresh fruit.

The Ministry of Agriculture survey shows that higher prices hit consumption of many foods. Beef prices rose 13 per cent and mutton and lamb 9 per cent, and the amount eaten per person dropped from 9.6 to 8.6 oz and 5.9 to 5.4 oz respectively.

During the last quarter of the year the number of eggs eaten per person dropped to the lowest level for five years: 4.5 a week each. Prices were 1½d a dozen up.

In spite of efforts to persuade people to drink more milk, the average consumption during the last quarter was the lowest for nearly a decade: 4.72 pints a person a week. But half the drop during the year was caused by the Government cancelling free school milk in secondary schools.

Bread sales continued to drop, and potato eating fell from 52.24 oz each a week in 1967 to 51.92 oz. (The Ministry said its figures exclude many Christmas food purchases and sweets, soft drinks and meals bought outside the home.)

The leader of Britain's £2,000 million-a-year processed food industry yesterday blamed the Government for food prices going up. Mr James Barker, president of the Food Manufacturers' Federation, told the federation's lunch in London that their prices were controlled by the force of free competition and that was why they found the Government's early warning system on impending prices unnecessary and undesirable.

A survey had shown that in the ten years before the early warning system, prices of manufactured foods rose 1.4 per cent a year – less than half the increase of foods in the retail food price index. The early warning system frustrated free enterprise and competition. It could cause manufacturers to be reluctant to cut the price of a particular product for fear that subsequently they would not be allowed to increase it if costs rose again.

Replying, Mr Cledwyn Hughes, Minister of Agriculture, said that the early warning system on food prices was an important part of the prices and incomes policy. The Government's aim was that, to the greatest possible extent, there should be a joint effort to develop a coherent and acceptable prices and incomes policy.

Christmas holiday food problem for wives – shops to shut for 4 days
Peter Bullen 8.11.69

Many food stores will close for four days this Christmas – the longest break they have ever had. They will shut on Christmas Eve, a Wednesday, and not open again until the following Monday.

The decision to stay shut on the Saturday will put an extra burden on housewives stocking up for their families. The Consumer Council said last night that it was "a very inconvenient decision" for shoppers.

Tesco, with 835 stores, said it was not opening on Saturday as it would be a waste of time. Sainsbury stores are also

staying closed on Saturday, but will open on Monday, a day when they are normally shut. A "large number" of Fine Fare's 1,200 stores and supermarkets will close for four days, and so will some Co-operative retail societies.

Most small grocers' shops should, however, stay open. Mr Len Reeves-Smith, general secretary of the National Grocers' Federation representing 20,000 family grocers and independent shops, said that all members were being advised to open on the Saturday. He said: "We believe the grocery retail stores have a duty to the consumer. The multiples are showing complete disregard for the public."

Small bakers and butchers are expected to follow the supermarkets and close for four days. The Master Bakers' Association said it was leaving the decision to individual members, but there was a general feeling that more shops would remain closed than would open.

The National Federation of Meat Traders' Associations also said it was leaving the decision to members.

Britain sprouts her own Brussels
Peter Bullen 24.11.69

The vegetable growers of Britain are preparing a move of patriotic defiance – the British sprout. No longer will the housewife need to pay court to the Brussels sprout, the vegetable that for centuries has flaunted its Continental pedigree.

It is thanks to 12 years of research by plant breeder Mr Alfred Ringer, 60, of Titchwell, Norfolk. "Our new variety is an all-British sprout," said Mr Ringer yesterday. "It has been bred for the housewife to meet her tastes and for our frozen food market." Until recently it was known only by its breeding number,

OS.20. From now on it is to be called the British-All-Rounder sprout.

Although other varieties had been bred in this country, most would have the Continental strains in them, said Mr Ringer. None would have been as British as his new variety. The All-Rounder is a darker green with tight small sprouts that are particularly crisp and sweet. Farmers would find them better to harvest as deleafing was easier and housewives would find they got better value for money with less wastage.

The first British sprouts will be sown in commercial quantities next spring and be on our tables in the autumn.

Rebel farmers may paralyse city traffic
Peter Bullen 15.12.69

Militant farmers may paralyse traffic in cities all over the country with slow-moving processions of tractors and farm machinery.

The warning was given yesterday by Mr Wallace Day, 52, Devon farmer and leader of the unofficial farmers' action group that caused an hour's snarl-up of traffic in Newton Abbot at the weekend. The hold-up, involving 40 to 50 farm machines, was staged to show farmers' desperate need for better prices from the Government at the next farm price review in February. Mr Day said: "Our party was entirely co-operative with the police. If we had wanted to, we could have bunged up the whole place."

An even bigger demonstration is being held tomorrow when 600 West Country farmers will march on Parliament to lobby MPs about their low incomes.

Mr Day said that since the action group's formation last month the response has been tremendous. People from other areas had phoned to say they wanted to form similar commando groups. Mr Day claimed that cities such as London, Birmingham, and Newcastle could be hit in one afternoon. "At this stage we have no intention of interfering with people's Christmas shopping. On the other hand if we get another bad price review, people can expect hold-ups in their Easter holiday traffic," he said.

Farmers had been forced to take militant action because, with the exception of beef, they were getting the same prices for their produce as they did ten to 15 years ago. Consumers did not realise this as they were paying more for their food. But farmers who had faced steadily rising bills for materials, machinery and wages, plus higher interest rates, were getting no more.

Farmers demand £140 million pay rise
John Winter 19.12.69

Farmers jumped into the queue for higher pay last night with an unprecedented demand for a £140 million rise in their incomes next year. The National Farmers' Union told the Government that unless this claim is met thousands of farmers will go bankrupt and home food production will decline.

If granted, it could mean an increase of £200 million in the total value of guarantees and subsidies at the February farm price review. This is almost four times the highest figure awarded at this annual settlement with the Government.

The demand came from the union's council of 146 members within 48 hours of a march on Whitehall by 500 Devon farmers and a mass meeting of 800 farmers in Warwickshire. But Mr Bill Williams, NFU president, who is fighting to hold his job as the farmers' leader, said the claim was being prepared before the demonstrations and petitions started.

Only once before, in 1964, has the union stated its review target figure. Mr Williams said: "I am stating our proposals now so that the Government knows well in advance where the union stands." He said it was impossible to give a precise figure until accounts of this year's costs and income are completed.

The claim is for a rise in the total net income figure of £509 million this year to £650 million in 1970–71. This increase would have to be enlarged by the rise in farm wages, bank interest and other costs.

The Government could meet the claim completely by raising the price review award or partially by import controls which would result in farmers getting higher market prices. This would raise food prices but, says the union, by not more than 1 per cent.

Mr Williams said: "This is the greatest crisis in farming since World War II. Farmers have neither the confidence to invest nor the cash to do it. But more money put into agriculture will increase the size of the cake for everyone."

Farmers 'are hardest hit by inflation'   *Peter Bullen 31.12.69*

Farmers have been worse hit by inflation than any other major sector of the community, Mr Asher Winegarten, National Farmers' Union chief economist, said yesterday. They are frustrated and angry because they have not shared in the general improvement in living standards in the past decade, he told Berkshire farmers at Reading.

"In the ten years up to 1968–69 the real value of total personal incomes in the community as a whole has increased by 46 per cent. Farm incomes, on the other hand, in real terms have shown only a 7 per cent improvement." Compared with the mid-1950s, agricultural prices including deficiency payments by the Government are only 6 per cent higher but prices of manufactured products have gone up 30 per cent, and retail prices nearly 45 per cent.

In the ten years from 1959 the following price changes had taken place:

- *Milk:* Farmers' price up 11½ per cent but retail price up 31 per cent.
- *Bread:* Farmers' price for wheat down 2 per cent; bread prices up 108 per cent.
- *Barley:* Farmers' prices down 13 per cent but draught beer prices up 63 per cent, excluding duty.
- *Beef:* Producers' prices up 28 per cent but retail price of boneless sirloin up 46 per cent.
- *Lamb:* Farmers' returns up 7 per cent, leg of lamb retail price up 47 per cent.
- *Pork:* Producers' returns up 5 per cent, leg of pork price up 26 per cent.

Over the same decade the labour cost per worker to farmers has risen by 75 per cent and farm rents by more than 100 per cent. In the past four years alone costs have risen by 17 to 18 per cent – double the increase that has taken place in agricultural prices.

Farmers must be allowed to improve their real incomes in line with their increased output and productivity he said.

For the Government to ignore this would mean ignoring the considerations of elementary social justice and the continued exploitation of farmers.

CROPS and MACHINERY

Ploughing may be going out

Peter Bullen 14.2.69

More than 8,000 million cubic yards of soil are moved every year by farmers to sow their crops – enough to build an earthwork 400 yards wide and 20 yards high from London to Inverness. But much of this huge operation may no longer be necessary.

Mr Jim Elliott, Agricultural Research Council weed-control chief, told the National Power Farming Conference at Brighton yesterday that chemical herbicides provide a new and powerful substitute for one of the main purposes of tillage – controlling weeds. Weed control by chemicals allowed freedom for a fresh approach to the whole subject.

Saturated land brings crops crisis

John Winter 19.3.69

Hold-up of spring work on rain-sodden farms has reached crisis point. The backlog of work throughout the country is now so immense that farmers have little chance of recouping this year the £40 million lost by last year's bad harvest. The acreage of winter wheat, 20 per cent down on December, is believed to have fallen still further behind. It is now too late for farmers to retrieve the lost acreage, in spite of the price increase expected in today's price review award.

Little more than half the number of early potatoes have been planted in the main growing areas. Sowing of sugar beet and later potatoes has barely started. Yet it is less than three weeks from the date when farmers have usually completed this work. The Potato Marketing Board yesterday reported that in Cornwall only 2,500 acres of early potatoes are in the ground compared with 4,000 acres a year ago. Pembrokeshire has 5,000 against 8,400 acres, and in the slightly later land of Kent the total is only 570 acres (5,000 acres).

Milk yields are down everywhere and farmers are having trouble with the level of non-fatty solids in the milk.

In the corn belt of East Anglia there have been only one or two brief periods when the land has been fit for machines. Work on all crops is seriously behind. Sowing of the region's 300,000 acres of sugar beet, two-thirds of the national crop, has not even started. Last year 80 per cent was sown by the end of March and farmers estimate that crops sown after April 7 are likely to yield half a ton an acre less for each week's delay.

The same rate of loss can be expected from potatoes. Yield losses of up to a cwt an acre a week are also feared from delayed cereals.

Grain prospects worst for years

John Winter 2.4.69

Cereal growers face the most disastrous two-year period in memory. Losses on this year's grain harvest, because of a cold, wet spring, will add £20–£30 million to last year's £40 million lost because of bad harvest weather. The gloom that has been building up among farmers as their spring cultivations and seed drilling fall further and further behind schedule is shared by Ministry of Agriculture advisory officers. They believe that serious yield losses are an inevitable consequence.

An official at advisory service headquarters said: "There have been spurts of furious activity on farms in the south and east on dry days in the past week, but by-and-large, work has been so delayed by the wet land that we cannot hope to recover from late sowing. On light land, given normal weather from now on, low yields are likely through the soil drying out. On heavier land, which has never had a real chance to dry out for two years, we cannot expect high yields."

It is 22 years since spring work was so seriously delayed – by the great freeze-up of 1947, although the summer which followed was exceptionally warm and dry wheat yields were 20 per cent and barley 12 per cent below average.

Mr John S Hopkins, Nottinghamshire county advisory officer, said: "This is the worst spring I can remember. I cannot recall a year when everything, including grass, was so backward. In my experience a short growing season never gives high yields. I believe that on light land in this area we have already lost three to four hundredweight of cereals an acre."

March sunshine at Kew, 65 hours, was only 58 per cent of the average and the lowest since 1916. Temperatures everywhere were below normal, as much as 5.6 deg F in the worst areas.

Two-spanner tractor on way
Peter Bullen 7.4.69

Work on machinery down on the farm will be simplified if a suggestion to the manufacturers is taken up.

When the farm machinery industry goes metric it is recommended that the sizes of nuts and bolts constantly needing adjusting in the field should be reduced to only four standard sizes. At present there is a host of different sizes, requiring 14 or 15 spanners. Most farmers need two or three sets of English, American and Continental spanners, but in a few years two spanners could do the job.

Last year, on the suggestion of the British Standards Institution, the National Farmers' Union and the machinery industry's professional body, the Institution of Agricultural Engineers, set up a joint working party to sort out the problem.

Mr Colin Brutey, NFU machinery committee secretary and chairman of the working party, said yesterday that it will be four or five years before the standardised sizes can appear on new machines. He added: "If firms will adopt this recommendation when they are redesigning for the change to metric, it could save whole hours of delay and inconvenience, not to mention barked knuckles."

Poison chemical spraying is safer now
Peter Bullen 28.4.69

A British firm has developed a machine to make the spraying of poisonous chemicals far safer. It works on the principle of rotary atomisation of liquids used by the US Navy during the last war to hide warships under banks of artificial sea mist. The Turbair machine can direct millions of minute droplets of chemicals exactly where they are meant to go.

The amount of chemical used per acre can be cut by about 85 per cent. Although the droplets are minute, they are not so small that they drift on the neighbouring crops.

The manufacturers have also introduced ready-to-use spray mixtures that are clipped straight on to the machine and so do away with the need for an operator to mix concentrated chemicals with water before use. The chemicals, in an oil base, are sprayed through the machine which looks like an ordinary small electric air-cooling fan.

The firm test-marketed its first machine only two years ago for horticultural and garden use. Since then it has exported to growers in 43 countries and is soon to introduce a tractor-mounted model for farm-land spraying. A director, Mr Desmond Heath, said yesterday that whereas other systems used pounds of chemicals per acre, they used only ounces to achieve the same results. This meant far less contamination of the environment. He added: "It is the safest spray machine in the world by a large margin."

At present it is mainly used for spraying small commercial areas of crops but it is also being used in agriculture for disinfecting livestock vehicles, grain silos and other buildings.

Farms face worst year since 1947
John Winter 21.5.69

Farmers face the most precarious year for their crops since the season following the great 1947 freeze-up. The disastrous effects of the cold, wet, sunless spring have already ruled out any hope of expanding home cereal production and will set back for two years at least the Government's plan to save, through increased home production, £160 million a year on food imports by 1974.

Experts yesterday estimated that with normal conditions for the rest of the year the bill will be one million lambs lost; cereals one million to two million tons below last year's harvest which was itself 1.3 million tons down on the 1967 record; and potatoes up to one million

tons down, which would cause a critical shortage a year hence.

The critical situation was brought home yesterday at the first big outdoor demonstration of the year, the national grassland demonstration at the National Agricultural Centre, Stoneleigh Abbey, Warwickshire. Two days before it opened 1¾ ins of rain fell on the 200 acres of demonstration land.

While 15,000 farmers watched 500 machines labouring in wet, muddy fields to cut and process grass, Mr Bill Williams, NFU president, said: "There are literally thousands of farmers working in almost impossible conditions this year. In Lincolnshire they are talking of leaving fields fallow."

Fido the robot tractor goes it alone
John Winter 13.6.69

Fido the tractor planted an acre of phantom potatoes at the National Agricultural Centre, Stoneleigh Abbey, Warwickshire, yesterday. A group of VIPs stood at the edge of the field and gazed incredulously at the empty driver's seat and untouched steering wheel.

Inventor Mr Peter Finn-Kelcey of Eversley, Hampshire, was demonstrating the world's first driverless tractor system which will be offered to farmers of the world at the Royal Show on July 1. For a matter of £1,000 (for Fido) and £8 an acre (for fixing his tracking system) farmers can replace a £1,000-a-year tractor driver with a system which is claimed to work for at least 15 years.

Fido is the automatic farmer's boy who can plough and sow, spray and mow, and do anything else normally done by a manned tractor. Fido is guided by a grid of wires buried 18in to 2ft below the soil. A guide-head slides along the boom fixed like a front bumper. He can be set to work at the speed appropriate to the operation in hand and will follow the underground wires to produce furrows straighter than the most expert driver, while a trip-wire round the edge of the field causes him to raise or lower his implement at the right point.

He will work on untouched by hand either until his field runs out or he completes the job and is switched off automatically. A cowcatcher in front gives an instantaneous cut-out if it touches a straying animal or other obstruction and there are safety devices to cope with other emergencies.

Mr Finn-Kelcey, 55, said: "This is a practical tool to help farmers to meet the growing shortage of labour."

Farmers hit the £25 million harvest jackpot   *John Winter 23.9.69*

The long hot summer has given farmers a golden jackpot of £25 million from increased production of corn to help them recover from last year's harvest disaster. Merchants estimated yesterday that the total harvest of wheat, barley and oats will be at least 13½ million tons. This is one million tons more than last year but 500,000 tons below the 1967 record.

The corn merchants say that forecasts of cereal yields by the Ministry of Agriculture are too low. Allowing for marked variations between areas and even from farm to farm Dr Clare Burgess, president of the National Association of Corn and Agricultural Merchants, said yesterday that all cereals will average more than the Ministry estimates of 32.4 cwt an acre for wheat, 29.1 cwt barley and 29.8 cwt oats.

This should give total crops of about 3.5 million tons of wheat (in spite of a drop of 360,000 acres sown, due to bad weather), 8.5–8.75 million tons of barley (1968, 7,944,000) and more than 1.3 million tons of oats (1,128,000 tons).

Grain quality varies widely. There has been a brisk demand for top-grade barley for malting, but Dr Burgess warned farmers against holding large quantities of lower quality barley to sell for animal feed.

"Feeding barley is not coming forward to meet demand," she said.

Farm poisons control to be tightened   *John Winter 1.12.69*

The Ministry of Agriculture is planning further action to tighten up control of the use of poisonous chemicals on farms. The present voluntary scheme, under which manufacturers register new products after careful testing before marketing them, is to be made compulsory.

The Ministry said yesterday that legislation is being prepared. In the past few weeks the Ministry has already decided to ban cyclamates in food and to curb the use of antibiotic feed-additives. Restrictions on the use of DDT and other similar chemicals that leave persistent residues are expected to be announced before Christmas.

Under the present voluntary pesticides safety precautions scheme, manufacturers make tests lasting several years to find what effects any new product has on plants, wildlife and humans and what residues it leaves. This information is supplied to a Ministry committee of independent scientists which draws up safety margins covering the quantities and methods of application of the chemicals to ensure that any risks to wildlife or of poisonous residues are kept to a minimum.

Chemical manufacturers fully support the voluntary scheme and some have even urged that it should be made compulsory. Recently the scheme was brought into line with similar arrangements in eight other European countries. Denmark, Switzerland and all six Common Market countries are keeping to the same standards for assessing the safety of new poisonous sprays and monitoring the effects of those already in use.

Spy planes check potato farmers
Peter Bullen 11.12.69

A row broke out yesterday over the use of planes to spy on farmers. The Potato Marketing Board has used the planes to check if farmers are growing more potatoes than they should. The spying tactic was revealed by Mr James Rennie, chairman of the board at its annual meeting in London.

One board member, Mr Jack Merricks, said after the meeting that the use of planes to check what potato producers were doing on their own land was diabolical. He added: "It's worse than the things they are doing in Russia. I have opposed this bitterly at board meetings."

The board has to regulate the amount of land on which potatoes are grown each year to prevent a big surplus which would lose the industry millions of pounds or a shortage which would cost consumers millions in higher prices.

Mr Rennie discounted suggestions that farmers would resent the move as a "Big Brother is watching you" tactic. He explained that the experiment had shown that it could be cheaper than the present practice of sending two inspectors around the country by car to check a third of the 640,000 acres of potatoes. It was also a useful psychological move to have producers aware that the aerial surveys were being done as any small plane flying above their farm could be from the Potato Board, he said.

Air surveys have been carried out on an experimental basis in Pembrokeshire and Lincolnshire. The results have encouraged the board to have flights over other areas next year. So far no disciplinary action has been taken against any producer following the use of spotter planes.

But in the past year the board has made 16,132 checks on the ground to see if producers are keeping to the acreage quotas. The board took disciplinary action against 22 producers from 13 counties for understating their potato acreages and fined them sums ranging from £10 to £75 each.

LIVESTOCK and POULTRY

NFU man attacked over 'leak'
John Winter 8.2.69

Mr Henry Plumb, deputy-president of the National Farmers' Union, was attacked in a Parliamentary notice of motion put down yesterday following a "leak" of the findings of the Government foot-and-mouth disease inquiry committee, of which he is a member. In the Commons on Thursday Mr James Wellbeloved (Lab, Erith and Crayford) said that on several occasions Mr Plumb had "made statements which obviously show that he entered into the work of the committee with a prejudiced and biased view." After being reprimanded by the Speaker, Mr Wellbeloved asked leave to raise the matter on the adjournment.

Mr Fred Peart, Leader of the House, who, as Minister of Agriculture, set up the committee headed by the Duke of Northumberland, had earlier said he was anxious to see its report, but it was not yet ready.

Now Mr Wellbeloved has put in a notice of motion saying that the implication in the leak, carried by *Farmer and Stockbreeder* on Tuesday, that Mr Plumb, the farmers' representative, has won a partisan victory is "highly damaging to the standing of the Northumberland Committee and detracts from its report."

The National Federation of Retail Meat Traders has written to the committee expressing serious concern at the leak, which suggests the committee will advise the Government to ban, at one year's notice, all meat imports from countries where foot-and-mouth is endemic, except chilled meat from which bones have been removed. This is based on scientific evidence that the virus is more likely to persist in bone marrow than in most of the edible meat.

Ministry is silent over £ for £ bid
John Winter 13.2.69

Nearly 1,000 farmers who lost their herds early in last winter's food-and-mouth epidemic are awaiting anxiously the outcome of their last-minute attempt to get extra compensation from the Ministry of Agriculture. The Ministry had offered to match £ for £ all money up to £250,000 put up by the industry itself for supplementary payments to them.

Many months passed between the time they lost their herds and when they were allowed to replace them, and prices were then far higher than those at the time of slaughter on which their compensation was based. After strong representation by the National Farmers' Union, the Ministry accepted that early victims were entitled to extra money and Mr Bill Williams, NFU president, launched an appeal. It brought in less than £100,000, but before it closed on January 26 the farmers who stand to benefit stepped in.

They provided £150,000 from their own pockets to raise the industry's contribution to £250,000. They expected the Ministry to contribute an extra £150,000. Their cheque was received at the NFU London headquarters, before the closing date, January 26, but the farmers who raised it have heard nothing since.

The NFU and the Ministry refused last night to say anything about it.

Mr Norman Ellis, on whose farm at Nantmawr, near Oswestry, Shropshire, the epidemic started, is chairman of the organising committee which canvassed all the early victims in Shropshire, Cheshire and Welsh border counties. He said last night: "We can see no reason why our cheque should not be accepted. It fulfils the conditions of the Ministry offer. We would willingly have taken action earlier, but the president's appeal was continued until the end of the year. We are very grateful that other farmers responded and gave as much as they did."

An NFU official in the area said: "There was a remarkable response from farmers who were early victims and there would be very great bitterness at any suggestion that their cheque does not qualify for matching contribution from the Ministry."

Disease outbreaks fell in 1968

John Winter 18.2.69

Britain had 187 outbreaks of foot-and-mouth disease in 1968 compared with 2,210 in 1967, says a return published today by the agricultural departments.

There were 98 outbreaks of fowl pest last year compared with 198 in 1967, and 211 outbreaks of anthrax compared with 438. There were no outbreaks of swine fever in 1968 – the second clear year since 1878, when the disease was first notifiable.

Why promise was broken by Ministry
John Winter 19.2.69

The Ministry of Agriculture yesterday gave its reason for breaking its promise to pay £150,000 to help the worst-hit farmers in last winter's epidemic. It said it had previously told the National Farmers' Union that "self-benefiting contributions would not be within the spirit of the £ for £ scheme," under which the Ministry undertook to double any sum up to £250,000 raised in response to a union appeal.

When the appeal fund looked like closing at £100,000 the 1,000 farmers in Shropshire, Cheshire and neighbouring counties who stood to benefit raised the outstanding £150,000, mostly from bank overdrafts, and sent it to the NFU.

The union, which has been pleading with the Ministry to honour its word and pay in a matching £150,000, said yesterday that the donors stipulated that their cheque should be paid in only if the Government matched it with an equal amount. The cheque has now been returned to the organisers. The 20-man committee of the affected farmers could yesterday hardly credit the Ministry's explanation because:

1. Many small contributions to the £150,000 cheque came from people who are not potential beneficiaries.
2. Most, if not all, of the 1,000 who stand to benefit have already contributed, in a variety of ways, to the £100,000 raised by the appeal, which the Ministry has matched £ for £.

The Ministry said last night that the farmers who had escaped the epidemic and benefited from higher prices should contribute to help the early victims.

Act with honour, Minister is told
John Winter 14.4.69

The 1,000 worst-hit victims of the 1967–68 foot-and-mouth epidemic have put the Government on its honour to fulfil its offer of extra compensation. At the weekend they sent the following telegram to Mr Cledwyn Hughes, Minister of Agriculture.

"For nearly two months 1,000 foot-and-mouth victims have been waiting for your Ministry to pay its promised share to the Foot-and-Mouth Appeal Fund. Delay is causing serious distress and hardship to many farming families. Request you to act urgently and with honour by paying the Ministry's promised contribution in full."

The telegram was signed by Mr Norman Ellis, chairman of the committee of 20 representing the men in Shropshire, Cheshire and adjoining counties who lost their herds in the first four weeks of the epidemic. It follows the report in Farm Mail on Friday which revealed that the Treasury is jibbing at paying £150,000, under a pound-for-pound offer because it was subscribed by farmers who stand to benefit.

When the cheques were sent to the

appeal fund in January, the refusal of the Ministry to match the amount provoked such a storm of protest that the Prime Minister called for a report. In the seven weeks which followed, the farmers who contributed have faced growing financial embarrassment. Many of them were granted extended bank overdrafts on the understanding that the money, plus the Government's matching payment, would be quickly returned. Not only has the Government's contribution been withheld, but the committee has been unable to refund the farmers own subscriptions in case it was called to resubmit the £150,000.

Mr Hughes, who is anxious to pay, has the support of Mr Fred Peart, Lord Privy Seal, who initiated the scheme. They are opposed by Mr Jenkins, the Chancellor. The final decision has awaited Mr Wilson's return from holiday. His verdict is expected this week.

Ban on all 'danger' meat from South America
John Winter 2.5.69

All meat imports which threaten Britain with foot-and-mouth disease will be banned, from October 1. This decision was announced by the Government yesterday, four hours after publication of the Northumberland Committee's report on the 1967–68 epidemic. The committee found that the £150 million catastrophe was caused by infected lamb from Argentina.

It would like to see a total ban on all meat from South America, where the disease is endemic. But recognising that this would cause a sharp rise in meat prices as well as serious political repercussions, it put forward an alternative. This is that future imports from suspect countries, Argentina, Uruguay, Brazil and Chile, should be limited to beef from which all material likely to harbour foot-and-mouth virus has been removed before shipment. The danger elements are bones, offal and lymphatic glands. Offal, such as kidneys, would have to be made safe by processing if sent separately.

The present total ban on mutton, lamb, and pork from these countries would continue.

The Duke of Northumberland, who headed the eight-man inquiry team, said yesterday: "We believe, with one dissentient, that our recommendations will reduce almost to nothing the risk of ever again importing foot-and-mouth into Britain."

The Government has decided to accept the alternative proposals, with an October 1 deadline for the switch to boneless beef. It has also agreed to reduce the duty of 20 per cent on boneless beef, which, compared with the 3 per cent charged on beef carcasses, would cause prices to rocket.

Providing the Argentinians quickly gear their meat plants for "boning", the changeover should not dramatically affect shop prices. The Argentinians at present send us 500–600 tons of boned beef a week, more than one third of the total.

The Government has made it clear that it will stand no nonsense from South American meat exporters by banning all meat from Uruguay from June 14. This follows a critical report from British inspectors on the condition of the seven refrigerated meat factories in Uruguay. The ban, announced on Wednesday night, will mean the temporary loss of about 10,000 tons of Montevideo beef a year. This is only about one-tenth of the amount we get from Argentina, and represents three days beef consumption in Britain. Imports from Chile are already suspended for similar reasons.

The Government will be ready to lift the ban as soon as the Uruguayans put their factories in order.

The Northumberland Committee believes that the present policy of eradicating foot-and-mouth by slaughtering diseased animals and farm stock in contact with the disease, is the right one for Britain. But it urges immediate steps to allow for vaccination of all farm animals within an area prescribed by vets surrounding an initial outbreak.

£50,000 to £-for-£ men

John Winter 8.5.69

The three months fight by 1,000 early victims of the 1967/68 foot-and-mouth epidemic for an extra £150,000 compensation ended yesterday. The Government refused to accept that £150,000 raised by the victims themselves qualified under the £-for-£ scheme. Under this the Ministry of Agriculture promised to match contributions up to £250,000 raised by the industry for early victims.

But Mr Cledwyn Hughes, Minister of Agriculture, announced that he would make an ex-gratia payment of £50,000 to the 1,000 farmers. This is in addition to the £100,000 the Government is providing to match the £100,000 raised earlier by the National Farmers' Union appeal fund. In a Commons written reply, Mr Hughes said that information from a number of sources, including the NFU president, confirmed that the £150,000 had been raised by a special appeal to the 1,000 early victims themselves.

It was on the basis that each affected farmer contributing £1 would, in effect, get it back with another £1 from the Government. This would have turned the Government's contribution into a direct grant. "This would not be in accordance with the concept of a £1-for-£1 scheme and the NFU have been so informed."

Happy cows in little cubicles

Peter Bullen 25.6.69

A herd of 35 British Friesians walked into Britain's first cow trap yesterday and locked themselves into their own tubular steel cubicles.

In winter they will spend all day and all night locked in the cubicles, except for milking twice a day, but at present they stay out in the fields at Penywaun Farm, 800ft up in the Welsh hills at Bedwas, Monmouthshire. Each cubicle measures 3ft 9in wide by 5ft 1in long. The cows lock themselves in by knocking a swivel frame.

The farmer, 38-year-old Mr Owen Watts, claimed yesterday that far from being cruel the cow traps are kind to cows. He said that since he installed the traps last November the cows had never been so healthy and contented. The quality of their milk had gone up and the average daily output had increased by a gallon a cow.

Trapping the cows in their cubicles meant he could handle a far bigger herd by himself and he was increasing his herd up to 60 cows. It also enabled him to isolate any cows in need of veterinary treatment. He said the cows went willingly into cubicles where they spent the time eating or sleeping on synthetic rubber mattresses.

Disease animals are banned by Royal Show

John Winter 5.7.69

The Royal Show from next year will accept cattle only from herds which are officially registered as free from brucellosis, the insidious disease of dairy cows which can also be caught by humans. This decision was taken unanimously by the council yesterday as the show at Stoneleigh Abbey, notched up another success with 44,000 paying customers. It will give a big boost to the Government campaign against the disease which costs farmers £2 million a year and can cause prolonged illness in people.

At this week's show the cattle are separated into three sections, according to their progress in tests against the disease. The total entry is the lowest for 20 years, and this has been blamed on regular exhibitors whose herds are already accredited refusing to risk contact with cattle that are not.

Mr Christopher Dadd, the secretary, said: "We believe this decision will bring cattle entries back to their previous level and it must encourage more breeders to enter the scheme." Mr John Madge, secretary of the Ayrshire Cattle Society, said: "It is a progressive move which we strongly support. It will certainly increase Ayrshire entries."

New export orders for British Friesian cattle signed at the show bring the total up to nearly 2,000 animals worth nearly £500,000. The orders are a record for this breed.

Total attendance for the first three days, 103,613, was over 9,000 higher than last year.

New bid to end cattle disease

John Winter 20.11.69

Doctors have joined forces with vets to try to force the Government into action to wipe out brucellosis, the dairy cows' disease which can be passed on to humans. A joint delegation from the British Medical Association and British Veterinary Association will go to the Ministry of Agriculture to press for an early start on wiping out the disease area by area. The disease costs farmers £2 million a year.

People who get the disease either through contact with an infected cow, or drinking unpasteurised milk, suffer from long-continued weakness, sweating and pains in the head, muscles, joints and stomach, which can sometimes lead to fever, rigors and mental disturbance.

So far Government action has been limited to encouraging farmers to clear the disease from their herds so that they qualify for a register of disease-free accredited herds. The BMA believes this actually spreads the disease, because it encourages farmers to get rid of infected cattle by selling them in the open market.

Mr John Parsons, BVA president, said: "We welcome the co-operation of the BMA. I believe this has got to be tackled by eradicating infected animals on a progressive area basis and that the incidence is already so low in certain areas that a start could be made now. Such areas are the south-western counties of England and parts of Wales." Eradication would mean the slaughter of animals and payment of compensation.

The Ministry said: "Apart from the cost of implementing a slaughter policy at this stage, there are many other factors pointing to the need for a gradual approach towards eventual eradication."

Ban on farm animal drugs

Peter Bullen 21.11.69

The Government is to curb the use of antibiotics on farms because they are a health hazard. The announcement came after publication of an inquiry committee report that linked the use of antibiotics on farms to the deaths of six people between 1964 and 1966.

Widespread use of antibiotics in animal feeds and for veterinary uses has created germs which cannot be treated with antibiotics, said the report. These germs can be passed on to humans mainly in meat and meat products. The germs in humans can then cause disease which is difficult or impossible to treat with antibiotics. The six people who died were among 60 cases of food poisoning in East Anglia. Their treatment was hindered because they had in some way picked up drug-resistant germs.

An investigation of the food poisoning outbreak by public health laboratories eventually found that the drug-resistant germs were identical to bacteria taken from calves. The calves that had been the source of the poisoning outbreaks had been given antibiotics for health reasons.

But the committee, headed by Professor Michael Swann, principal of Edinburgh University, was mainly concerned with the use of antibiotics put in animal feeds to boost growth. Because of the hazard this created by preventing proper medical treatment with antibiotics, the committee recommended stringent curbs on the use of the drugs on farms.

The recommendations have been accepted by the Government. Farmers will not be able to buy and feed penicillin, the tetracyclines and other antibiotics to pigs and poultry. In future these drugs will be available only on a veterinary surgeon's prescription and then only to help animals recover from illness.

Alternative safer feed antibiotics will eventually be substituted. They will not need a veterinary prescription.

 # EMPLOYMENT

Give our men that 17 shillings say farmers
Peter Bullen 4.1.69

Farmers all over the country are spontaneously backing the farmworkers' call to the Government to leave their 17s-a-week pay increase intact. Most of the criticism is being levelled at Mrs Barbara Castle, Minister for Employment and Productivity, who has referred farmworkers' pay to the Prices and Incomes Board for investigation.

In the West Riding of Yorkshire, members of the NFU executive expressed dismay and abhorrence at the Government's attitude to the workers' award, said the county secretary, Mr Frank Marsden, after the meeting.

Mr Bill Williams, National Farmers' Union president, told the Cumberland and North Westmorland branch annual meeting yesterday that the NFU regretted the action the Government had taken. Some members of Kent NFU executive are also critical of the Government's action. East Sussex county executive has passed a resolution deploring the reference of the workers' pay to the PIB.

Row as Reds gate-crash rally
John Winter 4.1.69

Communist banner holders refused repeated requests to leave the speakers' platform at the annual Tolpuddle Martyrs memorial rally yesterday. More than 2,000 trade unionists, mostly farm workers, thronged the little village of Tolpuddle, Dorset, to remember the six farm workers deported to Australia in 1834 for daring to set up a trade union branch. Several times organisers appealed over the loudspeakers for the others to leave the platform but six groups refused to move.

Later Mr George Woodcock, former TUC general secretary, said if it were ever necessary to have legislation to curtail trade unions it could be only as part of a new comprehensive labour code which bound the employers as much as the unions.

Mr Bert Hazell, MP for North Norfolk and president of the National Union of Agricultural and Allied Workers, announced that the union had sent its new claim for an increase of minimum farm wage from £12 6s to £16 and a cut in the working week from 44 to 40 hours for vetting by the TUC's incomes policy committee.

Farmworkers get 17 shillings

Peter Bullen 21.1.69

Farmworkers are to get their 17s a week (7⅓ per cent) pay rise next Monday after all.

Earlier this month Mrs Barbara Castle, Productivity Minister, angered both workers and farmers by insisting that the Prices and Incomes Board should investigate the increase because it was more than the 3½ per cent maximum allowed under the Government's incomes policy. But the PIB reported yesterday that farmworkers were the lowest-paid major group of workers in the country and deserved the rise which will cost farmers about £14 million a year.

The PIB could not approve the increase itself as it did not meet the strict criteria on productivity set out in the Government's White Paper on incomes policy. So the PIB passed the decision back to Mrs Castle by recommending that the Government should make a special exception of farmworkers' pay. Then Mrs Castle announced in Parliament yesterday that in light of the report the Government did not propose to take any action to delay the implementation of the Agricultural Wages Board award.

In its report the PIB said that on grounds of low pay alone, farmworkers should qualify for an increase. Average weekly earnings in agriculture were £15 8s 3d compared with £22 5s 3d for men in other industries. The gap in hourly earnings was even greater: 3s 3d an hour less than the average industrial rate. Farmworkers also worked about three hours longer than industrial workers each week.

Both sides of the industry contended that the 17s a week extra could be offset by the workers' contribution to productivity improvement. The increase will bring the weekly minimum rate up to £12 8s for a 44-hour week.

Levy threat to rebel farmers

Peter Bullen 11.2.69

Mrs Barbara Castle's department got tough with farmers yesterday over their long-standing disagreement about the Agricultural Training Board. Leaders of the National Farmers' Union and representatives of the Farmworkers' Union and the board went to the Department of Employment and Productivity to hear the Parliamentary Secretary, Mr Roy Hattersley, declare that the board was here to stay, and that farmers must make up their minds quickly how it should be financed.

Mr Hattersley said later that he had instructed the board to start taking to court the 26,000 farmers who still owe their £3 a worker levy for last year (total arrears £250,000). The department was also empowering the board to start sending out its levy demands of £3 10s a worker for the current year to the 93,000 employers.

Mr Hattersley said the Government had rejected the two proposals by the NFU for financing the board: either from a deduction from farmers' SET refunds or from a block contribution out of the price review award. But the Government was prepared to finance the board through a deduction from the annual £30 million fertiliser subsidy. Farmers could choose either that system or continue paying the levy direct to the board.

Equal pay worry for market gardens   *Peter Bullen 24.2.69*

Equal pay for women working on the land is strongly opposed by many employers, especially market gardeners, a survey by the National Farmers' Union labour committee shows. Minimum weekly wage for women on the land is £9 6s for 44 hours – £3 2s below the men's rate.

When the union heard that Mrs Barbara Castle, the Productivity Minister, had said the Government's long-term aim was equal pay for women, the labour committee decided on a survey. The flowers committee reported that the idea "immediately brought very strong protests from members."

The committee added that since market gardeners employed far more women than farmers, the proposals would weigh heavily against the market gardeners, who would not be able to make up their losses through the Farm Price Review procedure. Women played an important part in flower seed production and equal pay would not only make the crop unprofitable but could stop production completely.

Four ways to beat the Training Board   *Peter Bullen 10.4.69*

A four-point campaign against the Agricultural Training Board was suggested yesterday. Mr Anthony Harrison, chairman of the farmers' anti-training levy committee, said:

1. The board's prosecutions of 27,000 farmers who did not pay the £3-a-worker levy last year could bog down in the slow process of law. It would take a year to deal with the undefended cases alone.
2. The committee would back a test case in which, he alleged, there were flaws in the board's claim.
3. Members representing employers could resign when their terms expired in August, and the National Farmers' Union could refuse to recommend replacements.
4. If the NFU council would not remove its representatives, county branches could suspend or expel them.

An NFU official said last night that he disagreed with Mr Harrison's theories on farmers' representatives. The Training Board estimated that three-quarters of the employers were paying the levy when summoned.

NFU gives final warning on training board   *John Winter 18.7.69*

The National Farmers' Union yesterday

gave the Government an ultimatum to clear up the three-year wrangle of the Agricultural Training Board. The national council in London decided to withdraw support from the board unless training arrangements satisfactory to the union are put forward by August 15.

This is the date when appointments of the board's original members end. The six NFU representatives have already decided to retire and the union will not nominate successors unless an acceptable formula has been agreed. The union's first demand is the scrapping of compulsory levies by which farmers are forced to finance the board with annual payments based on the number of workers they employ.

Non-payment of the 1967–68 levy of £3 a worker led to the first five farmers appearing in court at Bromley, Kent, on Wednesday, when all the cases were adjourned for a fortnight after seven hours of confused evidence and argument. The board has issued 28,000 summonses and it is estimated that about 13,500 farmers are prepared to sit tight until they are taken to court.

The alternative method of paying for the board, which has so far been rejected by the Department of Employment and Productivity, is with money taken as farm production grant out of the annual price review.

Letter of the week – we trained our postman

Jenny Gooding, Ockley, Surrey 28.7.69

The thing that does not encourage farmers about training in agriculture is the fact that trainees, taught the job, turn to factory jobs where there is a better wage. My husband trained a lad, taught him the lot – he now works as a postman.

End of the levy   *Peter Bullen 14.8.69*

Farmers gave a very cool reception yesterday to Mrs Barbara Castle's plan to end the three-year-old battle over the Agricultural Training Board. Despite winning the major concession from the Government that farmers will no longer have to pay the hated, compulsory levy to finance the Board, the National Farmers' Union emergency council decided to submit Mrs Castle's scheme to their county branches.

Mrs Castle had offered to replace the levy with a new method of paying for the Board. Every year, at the annual farm price review, the £1 million to £1.5 million needed for the Board will be deducted from the bill of increased costs that farmers ask the Government to meet in subsidies and grants. From next month – the start of the Board's new financial year – 95,000 farmers and horticulturists will be released from levy liability if they accept Mrs Castle's scheme. For horticulturists whose products do not come under the price review, the new scheme will mean they get the board's training facilities for nothing.

But many militant council members demanded that the current year's levy of £3 10s a worker – still owed by 69,000 farmers – should be included in the new scheme. Others called for a new chairman and new members of the board. Mrs Castle is more likely to accept this.

By financing the board out of the price review funds, savings of £150,000 to £250,000 a year will be made in administration and collection of levies. But Mrs Castle's Department of Employment and Productivity made it quite plain yesterday that if the new financing scheme is accepted, the board will then proceed with its plans for collecting outstanding levies for last year and 1968/69 in full. About 11,000 farmers still face court action over non-payment.

New deal for the tied cottage worker
John Winter 30.10.69

Farmworkers may have greater protection soon against being thrown out of tied cottages when they leave their jobs. A new Bill published yesterday allows in most cases six months' grace to a worker, from the end of his employment – or to a widow from the date of her husband's death. Any possession order must be suspended for that period. Farmers who secure possession of a cottage by misrepresentation, and then let it for other purposes, such as a holiday or weekend home, would be liable to a damages claim by the outgoing worker.

The Bill also proposes important changes in the Government scheme to encourage amalgamation of small, unprofitable farms into bigger units. At present an amalgamation farm has to remain undisturbed for 40 years. This has been a major reason for the disappointing number of applications. Only 734 schemes have gone through in two years. The Bill reduces the standstill period to 15 years.

Simplification of procedure to get Government grants for improvements will reduce form-filling. Applications for 11 different grants will be covered on one form. Mr Cledwyn Hughes, the Minister of Agriculture, told the Country Landowners' Association yesterday, that his aim is to streamline the grant system, and cut out unnecessary paper work.

He also hoped to encourage farmers to create more viable holdings and co-operate to secure better returns. But he added: "At the same time, let no one say

the family farm does not have a first-class future. It is of inestimable value to this country."

The Bill provides for improvements in Britain's 13,000 smallholdings, and the creation of an Eggs Authority, to take over responsibility for egg marketing. The authority will have 12 members, all appointed by Ministers.

Record rise for workers on the land
John Winter 6.11.69

Farmworkers last night won a record pay increase of 15s, plus a cut of one hour in their working week. The award, worth a total of 28s a week, was made after seven hours of arguments in the Agriculture Wages Board. It will take effect in February, just 12 months after the workers got their previous biggest rise of 17s, and a total of 350,000 workers in England and Wales will benefit.

The minimum for full-time men is increased to £13 3s for 43 hours. It raises average earnings to about £18 10s, but this leaves the farmworkers still about £6 10s below the average in industry with three hours longer to work for it.

Mr Bert Hazell, MP, president of the National Union of Agricultural Workers, said: "I feel this award is not sufficient to affect the loss of labour from the land (about 14,000 workers last year). We are disappointed that it falls so far below our target of £16 for a 40-hour week. But it is the biggest increase we have ever had and there is some satisfaction in at last breaking the 44-hour week with which we have been stuck for years."

Mr Henry Sharpley, chairman of the NFU labour committee, said: "It is bound to affect food prices. It must raise the price of commodities like poultry and horticultural produce, which are not covered by guarantees at the price review."

The farmers say the award will cost them £20 million in a full year.

NFU told: You gave Barbara blank cheque   *John Winter 12.11.69*

Farmers in revolt against the Agricultural Training Board yesterday accused NFU leaders of railroading the council vote in favour of a Government compromise solution to end the three-year training dispute. They called on members to replace Mr Bill Williams, the president.

Mr Anthony Harrison, chairman of the farmers' anti-training levy committee, said the decision – to fund the board from the price review instead of by levies on 92,000 farmers – was clearly contrary to most members' wishes. It would give Productivity Minister Mrs Barbara Castle a blank cheque to take an unlimited sum for the board.

"The part she has played in this sorry affair has been intolerant, inflexible and inglorious," he said. "Some £2 million of taxpayers' money has been quietly pumped into the board, £560,000 of it after her junior Minister promised Parliament that no more money would be forthcoming."

Mr Harrison, who farms at Steyning, said the West Sussex branch was 90 per cent certain to nominate Mr Henry Plumb, deputy president, to succeed Mr Williams in January.

Mr James Reedy, head of NFU information, said: 'We very much regret these baseless allegations. Farmers were able to state their views and the council came to a democratic decision. We do not consider that the Minister has a blank cheque. We have closer financial control than under the levy system."

Strike threat   *Peter Bullen 1.12.69*

Farmworkers' strikes may be used to back their wage claims in future. This was indicated by a speech by Miss Joan Maynard, vice-president of the National Union of Agricultural and Allied Workers, at Brighton. She said they were in a more militant mood than ever before after the "totally inadequate" pay award on November 5.

"Hampshire, at their recent county conference, passed a resolution for strike action. We must give serious attention to this matter; the farmers are particularly vulnerable at certain points and at certain times of the year."

Board's new bid to stir farmers
John Winter 2.12.69

The Agricultural Training Board has appealed for the co-operation of 53,000 farmers who have so far ignored the board. Letters have been sent to farmers who failed to respond to an earlier request, asking how many workers they employ and on what jobs. The board says it needs the information to plan training.

But it refuses to disclose how many farmers have responded to its ultimatum to pay levies of £3 10s a worker for 1968–69. On October 30 it posted notices to 65,000 farmers threatening legal action if they had not paid within 14 days. At the end of the period only 5,000 had paid and yesterday the board said it could not give any further figure until next week.

In the Appeal Court on Monday the board will appeal against judgements against it in Bromley County Court last July in two cases in which it sought payment of levies outstanding from 1967–68. More than 9,000 farmers who received summonses for that year's levies have still not paid.

Mr Anthony Harrison, Sussex farmer who leads a farmers' committee which is challenging the board, said: "The board is bankrupt, with liabilities of £298,000, and living on bridging loans of £50,000 a month from the Department of Employment and Productivity." Unpaid levies for the two years are thought to be about £750,000.

Farmworkers pick a union fighter
Peter Bullen 18.12.69

The National Union of Agricultural and Allied Workers last night elected Mr Reg Bottini, 53, as general secretary. Mr Bottini, head of the union's wages department for 15 years, will bring a new militancy to its fight to improve conditions for farmworkers.

He said the first problems he will tackle will be wages and tied cottages. Recent proposals by the Agricultural Wages Board to raise the minimum wage by 15s to £13 3s a week and reduce the 44-hour week by one hour, did nothing to bring farmworkers' earnings and conditions in line with those in other industries, he said. The union would not rest until it achieved parity and also removed the tied cottage and the dread farmworkers had of losing the roof over their heads.

Mr Bottini said a drive would be started to improve the membership of the union, which has 115,000 farmworkers affiliated to the TUC.

In the poll Mr Bottini received 19,964 votes, Mr Dennis Hodsdon, assistant general secretary, got 17,092 and Mr Edgar Pill, the union's education chief, had 7,783. Mr Bottini succeeds Lord Collison, who retired in September after 16 years as general secretary to become chairman of the Government's Supplementary Benefits Commission.

 PEOPLE in the NEWS

The 363 hours I worked for the Ministry
John Winter 9.1.69

A farm manager told the Oxford Farming Conference yesterday that last year 363 hours of management time on his seven farms were spent in seeing Ministry of Agriculture officials and filling in forms. Mr George Douglas, whose 750 acres produce milk, potatoes, barley, peas and daffodils, was criticising "bureaucratic ineptitude and interference with farmers' freedom".

Mr Douglas said that, on his own dairy farm, the Ministry refused a grant for a milking parlour which could be moved twice a year between summer and winter sites, but said that a grant would be payable on two permanent parlours at £2,000 apiece. "We were so amazed that we queried the logic of it, only to be informed that the decision stood," he said.

Another charge by Mr Douglas was that the Ministry had forbidden its senior agricultural advisory staff to attend the conference because of its political flavour. But a Ministry representative said: "Four advisory officers are attending the conference as official observers, and others have attended individual sessions."

Mr John Mackie, Parliamentary Secretary to the Ministry, said of Mr Douglas: "He spent half an hour in destruction, and only seven minutes in construction." Mr Douglas had raised the question of his parlour to suggest that the whole farm improvement scheme was a mass of bureaucracy. There had to be a system which prevented abuse. Mr Douglas had not mentioned the tens of thousands of improvement schemes which sailed through without any trouble.

Mail man wins the top farming award
Peter Bullen 21.1.69

The Fisons Award for outstanding agricultural journalism has been won by John Winter, agricultural correspondent of the *Daily Mail*. The award is in recognition of his reporting in the *Daily Mail* of Britain's worst foot-and-mouth disease epidemic a year ago.

In particular the judges singled out his feature 'The Silent Disaster' that described the anguish on one of the farms where all the animals were slaughtered. In making the award – a £250 travel scholarship – the judges praise the way he reported the epidemic, showing clearly the agony and dilemma of those farmers who had their businesses wrecked by the outbreak.

The Fisons Awards, administered by the Guild of Agricultural Journalists, are made annually for outstanding journalism in the technical and non-technical Press. John Winter received the award for the General Press from Lord Netherthorpe, chairman of Fisons, at a ceremony in the Press Club, London, last night.

The £250 Specialist Press award was shared by Michael Leyburn and Michael Williams of the *Farmers Weekly*, and Richard Knight, of Southern Television, won the £150 under-30s award.

The award confirms the *Daily Mail*'s unrivalled coverage of farming. In 1965, Peter Bullen, assistant agricultural correspondent, won the under-30s Fisons Award and last year he was commended for a feature on the agricultural implications of the Stansted Airport plan.

BBC says sorry over TV farm errors
Peter Bullen 5.6.69

The National Farmers' Union accused BBC TV last night of deliberately screening a biased programme on farmworkers' conditions. The programme, BBC 2's *Man Alive,* showed a number of farmworkers and their families living in tied cottages on the minimum £12 8*s* agricultural wage.

The NFU was upset that the programme had gone out under the heading *An Everyday Story of Countryfolk* when the union had a letter from Mr David Attenborough, BBC Director of TV programmes, saying the programme did not attempt to show well-off or average farmworkers' families.

Mr James Reedy, head of the NFU's information services, said last night: "The BBC has apologised to the union for errors in its research. The BBC has admitted that the programme did not show normal conditions and that it set out specifically to try to find families living on less than £13 a week. The BBC agreed last night that minimum earnings were nearer £15 a week and it asked the union 'to accept the BBC's sincerest apologies for what was a serious error on the part of the research team involved'."

After 40 years, 'Harold' leaves the land
John Winter 23.8.69

Lord Collison, known as Harold to thousands of farming men, is to resign as leader of the Farm Workers' Union. He will leave his job as general secretary on September 30 to become chairman of the Supplementary Benefits Commission.

Lord Collison, 60, started work on a Gloucestershire farm 40 years ago, and has led the farmworkers for 16 years. He was chairman of the TUC in 1964–65.

He said last night: "It was not an easy decision to make. It means a complete change from the work to which I have devoted most of my life. I shall never lose my interest in the land and the men who work on it."

Farmers' leaders to be groomed for TV
Peter Bullen 24.11.69

Farmers' leaders are to attend special television grooming courses in the New Year. The National Farmers' Union is now selecting its top office holders and officials for the treatment. Like many politicians who have had similar instruction, the NFU feels it is vital that its leaders are able to put over the industry's message well.

Mr James Reedy, the NFU's information chief, who has taken the course, said yesterday that sooner than employ television professionals to put over their case, they preferred to have their own farmer leaders learn the television techniques. Farmers were used to using the most modern machinery on their farms. He felt television was just another new technique they would soon master. It would be of particular benefit when farmers were confronted on television with sophisticated professionals.

The intensive, one-day courses run by Bath University of Technology show people how to get the best results. They are told how to handle difficult interrogation and how to deliver their case simply and briefly. Advice on mannerisms, appearance and behaviour is also given.

A Dorset farmer, Mr Robert Saunders, who is NFU publicity committee chairman, has attended the course. He said he found it a great help. He added: "Agriculture has a very strong case to put and it is highly desirable that those who have a part to play in putting it should do so as effectively as possible."

In the past, other industries, whose executives were city-based, had had the advantage over agriculture when it came to television appearances. Agriculture, by its nature, was composed of scattered individuals from rural areas who were not likely to appear frequently before television cameras.

GENERAL

Children who fast for 18 hours each day
Peter Bullen 24.1.69

A survey of 80 London schoolchildren aged ten and 11 showed that a quarter regularly went without food for 18 hours – from an evening snack to lunch next day. They had no breakfast.

Another study was made of 75 secondary schoolchildren. Of those who had school lunches 47 per cent habitually left foods untouched, invariably vegetables. Twenty children given money to get lunch outside school bought pie and mash or sausage and chips. Not one included green vegetables. In six cases only chips and sweets or a soft drink were bought.

These are the first results of an investigation into children's eating habits commissioned by the National Dairy Council. They are given today by Dr Ben Lynch, head of London University's nutrition research unit.

Writing in the *Medical Officer*, Dr Lynch questions the influence on nutrition of girls trying to qualify to wear micro- and mini-skirts. And what were the implications when the leading model was Twiggy with vital statistics of 32-21-32½ and her male counterpart was similarly slimly built? Dr Lynch said they had interviewed an 11-year-old girl who said she had not had breakfast because she was interested in slimming.

Long-haired, pale-faced teenage boys told him in a London club that they wanted to look "slightly ill" so that they looked "interesting".

Dr Lynch said many teachers were raising questions about fatigue creeping into school routine. Television viewing was being blamed. But he believed poor nutrition was the cause of some of it. The doctor estimates that if poor nutrition impairs the learning ability of only one in four secondary schoolchildren by 25 per cent between playtime and lunch it is wasting more than £5 million. He compares this with the £4 million cost of the free milk for secondary schoolchildren now scrapped by the Government.

Last night two Manchester MPs, Mr Alfred Morris and Mr Will Griffiths, both Labour, tabled questions asking what Mr Edward Short, Education Minister, intends to do about the unit's report.

Beauty spot farms get cash aid

Peter Bullen 5.9.69

Farmers in two beauty spots are to get grants to provide amenities on their farms for hikers, tourists and other countryside visitors in an experimental plan announced yesterday. Under the three-year scheme, the Countryside Commission will give farmers a unique cash-first-work-later grant offer "to enable them to do small tasks on their property which will aid the visitors' enjoyment."

If the scheme succeeds it may be extended to farms all over the country in areas of outstanding beauty.

Fairly small areas of 30,000 acres in Snowdonia and 23,000 acres in the Lake District National Parks have been chosen for the pilot scheme and £2,750 allotted. Agricultural Land Service officers will visit farms to work out what work is to be done and the size of the grant. Payment before the work begins is almost unprecedented, a Countryside Commission official said yesterday, but as farmers were being asked a favour it was only fair.

Payment might be made for marking footpaths, clearing litter, building picnic seats, repairing walls and fences damaged by visitors, planting trees and providing parking facilities.

Welcoming the scheme yesterday, Mr David Lloyd, National Farmers' Union information officer for Wales, said the big build-up of visitors to the countryside recently had caused concern. The NFU had pressed strongly for Government action to ensure that if townspeople were encouraged to visit the country, the farmer's "factory floor", his land should also be safeguarded. "The experiment is certainly a step in the right direction. But farmers will not welcome any move which might attempt to divert them from the priority task of feeding the nation and turn them into glorified part-time park keepers."

Seven million acres still undrained, says Minister

Peter Bullen 11.9.69

More than seven million acres of Britain's 27 million acres of farmland have no proper drainage systems, Mr Cledwyn Hughes, Minister of Agriculture, said yesterday. The industry's productivity could be improved by drainage, Mr Hughes said, when he opened the Black Sluice pumping station at Boston, Lincolnshire.

Under the £2 million scheme, more than 100,000 acres of Lincolnshire farmland have benefited from the construction of 24 pumping stations and improvements to 100 miles of waterways and drains in the Black Sluice area. The Minister said that 150,000 acres were now being drained, and there was every indication of a steady increase in the future.

At the same time, on a farm in Nottinghamshire, Boots the chemist was opening a new intensive dairy unit where the Minister's drive to improve farm drainage was being put into practice. The farm is part of the 1,000-acre Thurgarton estate, near Nottingham. Until this year, many of the fields relied on a drainage system developed through the labour and farming forethought of the monks of the original Thurgarton Priory, dating to the 11th century.

Today, the 280-acre dairy unit on the estate boasts an immense concrete and steel building costing £42,000 – after Ministry grants – that houses 200 milk-ing cows, calves, two bulls, food stores, automatic feed milling equipment and a milking parlour.

Farm manager Mr Harold Tyldesley said that until this year no drainage work had been needed. But the weather had been so bad in the 12 months up to this spring, that new drains were being put in for 35 acres and further work would have to be done.

When it is in full production the dairy unit, worked by three full-time stockmen and one part-time man, will be producing 180,000 gallons of quality Ayrshire milk annually. This should make the farm a net profit of £11,000 a year.

Concorde faces glasshouse test

Peter Bullen 10.10.69

Supersonic bangs simulated with explosives are to be aimed at glasshouses in tests to assess the likely effect of the Concorde airliner flying overhead. Government scientists are co-operating in the tests to be carried out at the Ministry of Agriculture's Efford Experimental Horticulture Station, near Lymington, Hampshire.

The Ministry said: "No aircraft will be used. We will be conducting carefully phased simulated sonic bangs with small explosive charges. They will be directed at certain glasshouses where instruments will record the effects. The tests are most unlikely to cause any disturbance to buildings or local residents."

The National Farmers' Union welcomed the tests. An official said: "We are concerned about sonic booms. Some growers have complained that their workers have been terrified to work in glasshouses after experiencing the vibration and noise of a boom."

Index

The Editorial Team

Peter Bullen

Peter Bullen's 40 years in journalism began on the *Surrey Advertiser*, Guildford, where he wrote the weekly "Gleaner" column of farm news. He moved to Lord Beaverbrook's *Farming Express* before joining John Winter on the *Daily Mail*. After the Mail and *Daily Sketch* merged he worked part time for the BBC's *Farming Today* programme before becoming Head of Public Relations for the Country Landowners' Association.

Several years as Deputy Commodities Editor of the *Financial Times* were followed by eight years as the owner of Past Delights Antiques in Guildford before he returned to journalism as Deputy Editor of *Agricultural Supply Industry*, Political Correspondent of *Farmers Weekly* and Political Editor of *Farmers Guardian*.

He is a Fellow, and former Chairman, of the Guild of Agricultural Journalists and is currently chairman of the GAJ's Charitable Trust. He has raised over £13,000 for the GAJ's Trust and other charities by running seven London Marathons, the last in 2002 at the age of 65.

Phillip Sheppy

Born in 1933 into a Worcestershire farming family, Phillip was educated at the King's School, Worcester and after national service returned to the family farm. He was elected Chairman of the Worcestershire Federation of Young Farmers' Clubs in 1958 and in 1960 was appointed Organising Secretary to the Northampton Town and County Federation of YFCs.

After spells with the Royal Counties Agricultural Society and The English Guernsey Cattle Society, Phillip was appointed Chief Executive of the National Proficiency Tests Council, a position he held from 1970 until his retirement in 1994. Appointed MBE in 1987, Phillip is a Fellow of the Royal Agricultural Societies and an Hon. Member of the City and Guilds of London Institute. He is currently Honorary Librarian to the Royal Agricultural Society of England.

Derek Watson

Derek Watson retired in 2000 after a career in agricultural journalism spanning nearly 50 years. Following a spell at a farm institute, he worked for *Farmer and Stock-Breeder* and *Farming Express* before joining Agripress Publicity in 1963. From then until his retirement he specialised in freelance journalism, contributing to a wide range of media. He wrote columns for several publications and was Editor of the Farmers' Club *Journal* for 23 years.

Derek is a former President and Treasurer of the Guild of Agricultural Journalists, and a Fellow. He was founder Secretary and Treasurer of the Guild's Charitable Trust. In 2001 the Guild presented him with the Netherthorpe Award for his outstanding contribution to agricultural communication and he was twice awarded silver medals by the International Federation of Agricultural Journalists.

Old Pond Publishing

Founded in 1998, Old Pond Publishing specialises in books and videos on farming and the countryside, with a particular interest in the recent history of farming and its machinery.

For further details and a free catalogue, please contact:

Old Pond Publishing,
104 Valley Road, Ipswich IP1 4PA, United Kingdom.
Phone 01473 210176
Fax 01473 220177
Email enquiries@oldpond.com
Website, with online ordering:
www.oldpond.com